Frederick William Cory

How to Foretell the Weather With the Pocket Spectroscope

Frederick William Cory

How to Foretell the Weather With the Pocket Spectroscope

ISBN/EAN: 9783744689007

Printed in Europe, USA, Canada, Australia, Japan

Cover: Foto ©berggeist007 / pixelio.de

More available books at **www.hansebooks.com**

HOW TO

FORETELL THE WEATHER

WITH THE

POCKET SPECTROSCOPE

BY

F. W. CORY,

M.R.C.S. ENG., F.R. MET. SOC., ETC.

WITH TEN ILLUSTRATIONS.

London

CHATTO AND WINDUS, PICCADILLY

1884

FORETELL THE WEATHER.

IN commencing an account of some new subject, it is customary to quote its history; but that of the rainband, one of the most important features in the sky spectrum, and the first noticed, has been so frequently referred to that it would be a matter of supererogation on my part to adduce what is now so generally known. Suffice it to say that in 1872 Prof. Piazzi Smyth, the Astronomer-Royal for Scotland, was the earliest to draw correct conclusions from its appearance.

The subject of which we are about to treat depends mainly upon the amount of invisible aqueous vapour contained in the atmosphere. Visible vapour influences the spectrum but slightly, and then only by reflected light. The records of hygrometry extend back for more than three hundred years. Mizaldus, in 1554, and Mersenne, in 1644, both noticed that the pitch of the strings of a violin varied according to the dryness of the air. The former says,* ' Musicorum instrumentorum subtensæ fidiculæ ruptim dissilientes, et hostia absque manifesta causa, aperiri claudique solito contumaciora, aerem pluvias nobis miscere palam nuntiant ;' literally meaning, ' The strained strings of musical instruments suddenly snapping, and sacrificed victims, without any apparent cause, opening and closing, and becoming more stiff than usual, plainly tell us that rains are disturbing the atmosphere.'

Even Pliny the elder, who lived A.D. 23 to A.D. 79, seems to have been a rough observer

---

* Mizaldus, A., ' Ephemerides Aeris Perpetuæ.'   Small 8vo.   Lutetiæ, 1554.

of the dew-point, as instanced in the following
quotation from the eighteenth book of his ' His-
toria Naturalis': ' Nec non in conviviis men-
sisque nostris vasa quibus esculentùm additur
sudorem repositoriis linquentia diras tempesta-
tes prænuntiant ;' which, translated, is : ' And
also at banquets and at our tables, vessels in
which there is any esculent moisture left on
the trays foretell fearful storms.' Of all the
numerous moisture-absorbers that man has
applied his ingenuity to in adapting to the
purposes of hygrometry, such as the beard of
the wild oat ; beard of musk grass, or geranium
moschatum ; internal membrane of the arundo
phragmites ; skin of the frog ; Dutch weather
house ; schistose stone, a material so porous
that when saturated it weighs nearly half as
much again as when dry; and then the more
modern and scientific instruments, viz.,
Daniell's, Regnault's, and Dines' hygrometers ;
and lastly, Mason's dry and wet bulb hygrometer, that is used more frequently than any

other at the present time (and which, by-the-bye, was invented by Boeckmann nine years before Dr. Mason was born), together with a great many others constructed of animal, vegetable, and mineral substances. Of all these, not one has the 'power of indicating the state of atmospheric moisture for more than a few yards around the instrument; but there is one that will inform you of the state of invisible moisture of a section of air many miles in thickness, the penetration depending on the amount of visible moisture present at the time of observation.

The following classification suggested by Pouillet is not perfect, but will do for all ordinary purposes.   He divides them into :

1st. Hygrometers* (or hygroscopes †) of absorption.

2nd. Hygrometers of condensation, *i.e.*, dew point instruments.

---

* ὑγρός (hugros), damp, and μέτρον (metron), measure.

† ὑγρός (hugros), damp, and σκοπέω (skopeo), to look at.

3rd. Hygrometers of evaporation, *i.e.*, dry
and wet bulb thermometers.

4th. Chemical hygrometers, for determining
the amount of vapour by analysis.
And I think we might now add to
these the hygro-spectroscope, or rainband spectroscope.

During the last few years this new form
of hygrometer has been rapidly advancing in
popular favour; but a want has been felt by
many an observer, which it will be my en-
deavour, in the following pages, to supply
in as brief, plainly worded, and exactly described manner, as will tend to assist the tyro
in his preliminary difficulties, with the manipulation and deductions therefrom, of an instrument of such fine and delicate research.

There are two spectroscopes at present in
use, both of moderate dispersion ; one, called
' The Rainband Spectroscope,' is an excellent
little instrument, very portable, and to a

skilled and careful observer reliable for forecasting.

FIG. 1.

The other, named 'Grace's Spectroscope,'
now becoming the favourite, is larger, but
with the same amount of dispersive power ;
the increased size of the spectrum in it is of

FIG. 2.—Spectroscope with attached Micrometer.

great utility in clearly defining lines and
bands, besides the advantage of being able to
use it earlier and later in the day than the
former kind, on account of the greater amount
of light admitted.

In several other respects it differs from the
first ; it has a milled wheel, with lever for

adjusting focus, and instead of plain glass nearest the eye, a concave lens is fitted.

Mr. Browning, the eminent optician in the Strand, has fixed to my Grace's an adjustable photographic micrometer, with prism of comparison. This combination makes the perfection of an instrument for meteorological purposes.

It would be as well now to proceed to a

FIG. 3.

description of the ' Rainband Spectroscope,' which, with the above exceptions, will equally apply to the other instrument.

It is composed of two cylinders, one fixed and the other movable, the latter sliding within the former; at the further end of the fixed portion, which is $3\frac{1}{2}$ inches in length, the aperture is covered with a circular piece of microscopical glass, and just within this are two parallel jaws, adjustable by means of a milled wheel outside, to admit more or less light, according to the require- ments at the time of observation. The movable part, $2\frac{1}{2}$ inches long, has at the extreme end a collimating lens to collect the light which passes through the slit, and to throw it in parallel rays upon the prisms placed

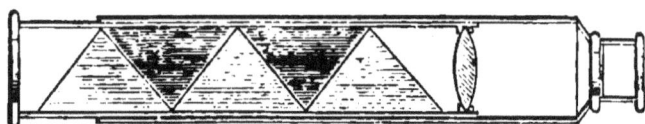

Fig. 4.—Section of Direct Vision Spectroscope.

behind it. These prisms are five in number : one, three, and five are made of crown-glass, two and four of flint-glass. (Vide Fig. 4.) They are all cemented together, and the effect of this combination is to give an image of the prismatic spectrum of sufficient dispersion to show the position of the dark lines, both constant or solar, and variable or telluric— that is, due to vapour in our atmosphere.

Between the prisms and the eye of the observer is another disc of glass. The total length of the instrument when closed is $3\frac{3}{4}$ inches, and its diameter $\frac{7}{8}$ of an inch ; while Grace's is $1\frac{1}{16}$ of an inch in diameter, and $5\frac{5}{8}$ inches long when closed.

In taking an observation with either instrument, shade the eyes with the hands in such a way that all extraneous light may be excluded as much as possible. It is of the highest importance to adjust the focus and slit of the spectroscope so that the lines in the spectrum may be of the clearest definition, and when once adjusted it seldom requires altering. I would recommend the beginner to focus either the D line in the orange part of the spectrum, or the E line in the green (vide Plate of Spectrum); and when he becomes more at home with the instrument, he will find no difficulty in rapidly and clearly defining any visible line in other parts of the prismatic spectrum.

The inner sliding-tube ought to be slightly

pushed in for examining the blue end, and drawn out a little for focussing lines in the red. To avoid loss of time in focussing, it is a good plan to mark the inner sliding cylinder with the edge of a knife, in the same manner as is frequently done to telescopes. If the slit be closed too much, horizontal lines parallel to the length of the spectrum will probably make their appearance; in almost every case they are due to the presence of dust on the edges of the slit. To remove this dust, open the slit as widely as possible, and wipe the edges with a small wedge of dry wood—an ordinary lucifer match cut into this shape will answer the purpose. Then close the slit completely; re-open it, and the lines will probably have disappeared ; if not, repeat the operation. Note that a camel-hair pencil, a leather, cloth, piece of paper, or blowing on it, will be sure to make the slit worse. Frequently, simply taking off the cap from the slit and moving the focussing tube in and out quickly will blow the small particles away. Sometimes, whilst observing, moisture by condensing on the eye-piece will somewhat obscure the spectrum ; to counteract this, warm the near end slightly either before the fire or sun, or by applying it to the warm skin.

In a general way, observe from ten to twenty degrees above the horizon, and towards the quarter from which the wind is blowing or likely to change to. As regards the latter point, it is much easier to forecast the probable direction of wind from movements of upper clouds, fluctuations of the barometer, etc., than it is to foretell rain.

The best hours for examining the general characteristics of the spectrum are 9 a.m. and 3 p.m., all the year round. In computing means, for various reasons it is not advisable to alter these hours, as by so doing errors are obviously liable to creep in. The focus and slit having been correctly adjusted, compare the lines and bands as seen in the spectrum with their relative positions as delineated on the chart (vide Plate I.), A, a, B, C, D, E, b, F, G, are constant and invariable solar lines ; a, B, c, c', α, r, δ, v, w, variable lines and bands ; B, α, δ, dry air lines, or low sun bands; a, c' r, v, w, rain bands ; r, position of principal rainband.

A, a, and B, are best studied by looking direct at the sun ; for in ordinary daylight B is not always well defined, a still less so, and A never, but, notwithstanding, is occasionally just visible. A is of no meteorological value, and changes very little for either high or low sun. It is not known what gas or metal in a gaseous state it represents in the sun ; this remark likewise applies to B. 'Neither of them are telluric lines, nor due to aqueous vapour, but must originate between the sun and our atmosphere' (Prof. Piazzi Smyth). The preliminary band of B is intensified with a low sun, and diminished with an increasing altitude. It is spoken of as a dry-air or low-

sun band. The physical origination of the
darkness of the band at a is due to an invisible
gaseous state of intensely watery vapour contained more or less in the lower part of the
earth's atmosphere (Smyth). A comparison
of the dark shadings at a and в ought always
to be made, if possible, when the sun is
shining at a low altitude. c is the red hydrogen line ; the lines about it with low sun are
probably due to water vapour. At c′ is situated a small rainband.

The α line must be produced by one of the
permanent gases of the earth's atmosphere.
The band at this position is telluric; a low-
sunband and of a dry-air character. Its
variations are of great importance in forecasting rain or fine, being very conspicuous
before dry weather, and on the other hand
weak in intensity previous to wet ; on one
occasion, when unusually distinct, fine and dry
weather prevailed over the whole of Western
Europe (Prof. Smyth).

Next in order is the principal rainband at
r, situated on the red side of the D lines,
which I will subsequently refer to more in
detail. In the yellow strip on the other side
of D a line is frequently seen when the spectrum at this part is unusually clear; consequently, before snow, when the vapour lines
have been considerably weakened from the
transformation of the invisible vapour into
snow crystals, this line (really made up of
several lines, including barium and calcium)
will be very prominent, at times exceeding in
darkness the lines in the rainband. The dry-
air band at δ is a marked feature when the
sun is low, and merges more or less into the
moist air shading at v, and on the other hand,
should the latter be strong, for instance, before a thunderstorm, it will be found to en-
croach on the shading at δ, a remarkable
phenomenon, the rationale of which is rather
difficult to satisfactorily explain. The line E
is one of the many that represent iron in the

sun. The *b* lines indicate the presence of magnesium, nickel and iron. The distinctness and definition of these important dark lines are of some value in the prevision of weather, as before heavy rain they are more or less involved in the darkness that envelops the violet end and almost obliterates the blue hydrogen line, F, in the glaucous, or sea-green coloured, region of the spectrum. The G group can be generally seen in ordinary daylight, and is a useful guide as to the clearness of the spectrum; this is the last set visible in small spectroscopes, but in Grace's, on several occasions of very clear atmosphere with a bright sun, I have succeeded in sighting the H lines in the lavender. Many persons mistake the G group for the H, which cannot be distinctly defined, even in large instruments of the prism variety.

THE RAINBAND.

The principal rainband is situated on the red side of the D lines, involving them, and

2

at times increasing or decreasing in intensity, and approaching or receding from the c line, according to the nearness or quantity of rain. In spectroscopes of small dispersion it exhibits itself as a dark shading; in larger instruments this band will be split up into a quantity of fine black lines (vide Plate II.).

Observers who have closely watched this band will no doubt have noticed great variations in its appearance—at one time being narrow and condensed, or broad and extended; at other times almost uniform in darkness, or rapidly shading off, or exhibiting itself as two dark lines to the red side of D, with one bright interval frequently between them, and another in the space between the more refrangible towards the green and D (Grace's spectroscope shows these lines exceedingly well). When they are strong and well-marked, they may certainly be taken as indicative of heavy rain. A thickness and darkness about the D lines alone must not be considered as a reliable rain prognostic. This illusory appearance leads people frequently to mistake it for the rainband proper, and to give it a greater value for rain than it sufficiently warrants. Undoubtedly, when the D lines are thickened, invisible vapour is the cause of it; but the two sets of lines on the red side of them are those to take into account in prevising rain.

In a great many anti-cyclones I have noticed this false band to be present. The rainband in many cases gradually increases for some time before rain. If it remains persistently high for several days, without rain falling in proportion, as much as a week perhaps of wet weather will follow. An increase in the percentage for the season of the year is suspicious of wet: in summer or warmer weather the rainband is higher than in winter or cooler weather; for example, 40 per cent. has a greater value for rain in the latter than in the former. Most of the

2—2

cyclones or depressions that cross England
from the Atlantic bring with them a large
body of air saturated with moisture. It is
interesting to gauge their individual proportion of rainband: a small number have 70 or
80 per cent., a very few only 20 or 30, but
generally 50 or 60. In the rear of a depres-
sion, as a rule, periodical showers occur ; the
air being cleared from the previous gale or
heavy rain, they can frequently be seen for
several hours beforehand with the spectroscope, but not with the naked eye.

Sometimes, during rain, the amount of
rainband may be low: this for the most part
denotes finer weather to follow ; if, on the
other hand, strong, it merely shows either
that more rain is to follow, or the spectroscope
is unable to penetrate the rain then falling
and analyze the light on the clear side of the
shower. Should there be a very heavy mist
at the time of observing, its telescopic power
will not be of much value. It would be too

much to expect of a spectroscope to sift the
light of several miles of visible aqueous
vapour; in such cases it is more on a par with
the hygrometers, that are only serviceable for
a few surrounding yards of rain. Intensity
of rainband does not presage the fall of snow;
on the contrary, the reverse does.

Mr. Capron, in his widely-known and
excellent little pamphlet, 'A Plea for the
Rainband,' has stated that the rainband is
low during cold winds. I can corroborate
this statement; and would add, that generally
speaking the rainband gradually diminishes
for several days before snow. It seems that
the vapour is transformed into snow-crystals;
and I have no hesitation in saying that in those
cases, when with a low percentage of rainband,
rain apparently falls, it is either melted snow
or hail; which is proved by the fact that
frequently, as the precipitation increases, the
rain decreases and only snow or hail is seen—
for in these cases it is highly probable that

the snow becomes melted in its passage through a warmer substratum of air, and as the temperature of this is gradually lowered by the snow, so we find an increase of the latter and a diminution of the rain.

If the spectroscope be directed towards the point from which the wind is blowing, and should the clouds be passing in one and the same direction, and 20 per cent. or less of the rainband be shown, no rain will follow for at least six hours, in spite of any threatening appearance of the sky.

In estimating low percentages of rainband, it is advisable to look direct at the yellow, so as to see the band slightly askew; by doing so any shading in that situation will easily become perceptible. Through observers not following this simple rule, one can quite understand how it is that frequent cases of 'no rainband visible' are entered on their charts or in their note-books. Such an absence of vapour lines seldom occurs, to my

knowledge. On one occasion that Professor
Smyth observed a disappearance of all the
rainband lines, a period of cold and dry
weather succeeded. It has never been my
good fortune to note a total absence of rainband; twice or three times the shading has
been so slight as to be barely recognised.
In the other extreme of percentage—for
example, where 80 per cent. is seen, the
darkness of the band is so extremely well-
marked as to form a most conspicuous feature
in the spectrum; should such an amount be
visible at the zenith, rain will certainly fall
heavily before long. Possibly every observer,
without exception, may at times be rather at
fault in the calculation of the quantity of
rainband. The use of a mental scale, which
one must of necessity have recourse to, cannot
be considered infallible. To compare the
darkness of the rainband with the permanent
lines of the spectrum will prove impracticable
if Grace's spectroscope be used; it might do

for instruments of very small dispersion, but
these are not to be recommended as hygro-
spectroscopes. As far as I know, Grace's
is decidedly the best instrument for this
purpose; and fitted up as I have already
mentioned, with a prism of comparison and
micrometer, the meteorologist may feel assured
he is possessed of an apparatus of immense
practical value.

Mr. Rand Capron, in enumeration of the
darkness of the rainband for the purpose of
record, uses a scale of from 1 to 5, as follows:
1 means faint; 2, faint to moderate; 3,
moderate; 4, moderate to strong; 5, strong.
Professor Smyth recommends and employs
from 1 to 10—which I adopted for some time,
until I could recognise intermediate shading,
and have now for more than a year found a
division of the scale into 20 parts as much as
it is possible to accomplish. A difficulty will
sometimes be experienced in ascertaining the
percentage at the zenith in certain atmospheric

conditions when the sky blue is very deep and dark; it certainly is not an easy matter, and all the careful manipulation and perspicacity at the command of the observer is requisite. It must be borne in mind that the rainband is quite independent of peculiar forms and characteristics of clouds indicating fine weather or the reverse.

Many cases have happened where a strong rainband has been noticed on a blue sky, with every appearance of fine weather for at least twelve hours, and rain has subsequently fallen within that time; or a very small percentage has been seen with most threatening clouds, that seemed as if they must shortly pour forth torrents of rain, yet it did not occur. Showers of rain may be very localized, and have a tendency to take certain routes, passing perhaps within a short distance of an observer, who, from the large amount of rainband visible at that time in the spectroscope, may prognosticate the rain that never reaches him;

and unwisely conclude, because it does not
fall in his immediate locality, that therefore the
instrument is worthless as an aid to forecasting. Such reasoning is clearly fallacious.
At first, when I was endeavouring to prove to
my own satisfaction the connection between
rainfall and rainband, I could not help feeling
disappointed at what I considered absolute
failures as regards the value of this band as a
fore-runner of rain; but, thanks to the information given by friends, reports in papers,
and by searching for evidence of rain having
fallen in the neighbouring country, with the
help, I ought to add, of that modern invention
of proved utility, the tricycle, I was enabled to
entirely put aside these misgivings, and rest
satisfied that what had been claimed for the
spectroscope was no mere chimera of the
imagination, but the commencement of a fresh
branch of meteorology that would eventually
supply the missing link in forecasting,
annihilate its difficulties, and bring to our

knowledge facts about the upper regions of
the atmosphere that were never dreamt of in
our philosophy.

Again, it is possible for a cloud to deposit
its shower before reaching the observer; and
by his following the rain-cloud with his
spectroscope he will in some cases become
sensible of a diminution in the rainband after
it has passed, and perhaps perceive what has
astonished me in one or two instances, and
that is, an actual increase in the band as the
cloud recedes. What can be the cause of this
increase? Is it an aggregation of nimbus
clouds? or is it perhaps a greater thickness
of the rain vapour as seen through the long
axis of the cloud? or is it possibly an alteration in the invisible vapour preparatory to a
heavier precipitation? These are questions
that are very difficult for a single observer
to reply to. The only way to solve this problem is to have reports from accurate spec-
troscopists, with numerous rain-gauges sta-

tioned at various points in the surrounding country.

Many a time, whilst watching waterspouts in the tropics, I have noticed that when one increases another decreases, although a distance of several miles may separate them. This fact may probably have some bearing on the question.

The more the correlation between rainfall and rainband is considered, so much the greater seems the influence of electricity and temperature in governing the precipitation of aqueous vapour; the latter, no doubt, is a great factor, for it has frequently been remarked that a sudden reduction of temperature has produced torrents of rain with a moderately high rainband.

Here the barometer is a most reliable guide, by informing us of the increasing body of wind prevailing over the ground current, which upper current may possibly be a cold north-east or warm south-west, the glass falling with the latter and rising with the former. Another item of especial value in predicting rain is to take account of the difference between the rainband at the zenith and near the horizon; the greater the range the less likelihood of rain, and *vice versâ*, except in cases of low percentage. Of course, if a thick mist is prevailing, the rainband may be the same wherever you point the spectroscope; and by noticing when it lessens at the zenith, you can foretell the clearing away of the visible aqueous particles. The value for rain in the last instance depends upon the amount of rainband present.

In several mists during the last winter both Mr. Rand Capron and myself have remarked the band unusually high. The indications of the rainband in all cases will be somewhat negatived if the dry-air bands are in excess, especially the α band. I would suggest to observers, for the better comparison of rain or snow and rainband, to take frequent notes of the temperature *in the rain gauge* of the rain that falls.

We will pass on now to the spectroscopic previsions of thunderstorms. The effect of these on the spectrum is to darken certain parts—for instance, the blue becomes deeper, and looks further away, the F line very indistinct, and the green apparently prominent and nearer the eye—a peculiarity I partly attribute to the increase of the rainband at w. This prominence of the green is only to be seen in spectroscopes of moderate dispersion, and not in the larger instruments. The blended bands at δ and v are much more conspicuous, and extend more towards the blue (they must not be confused with the dark band seen at δ with low sun). The rainband may be much or little, depending greatly upon the quantity of hail and rain attending the storm. This hail in a few instances seems to veil the rain-vapour behind

it from the spectroscope, and proves a dis-
appointment to the observer who depends
entirely upon the rainband. The small
amount of the dry-air band at α must also be
considered, and an increase of the a with less
distinctness of the b. All these features make
up the special spectrum noticeable more or
less in advance of thunderstorms.

I well recollect one instance in which all
these features were so very strongly marked
at the time of observing, about 10 a.m., that
I confidently prophesied a storm would take
place, although none of the other instruments,
nor any particular aspect of the sky, nor high
temperature with oppressiveness, indicated
that such an aërial disturbance was about to
occur; yet certain enough, in the evening,
about seven o'clock, the storm did come, much
to the astonishment of a great many, including the friends with whom I was staying.

The prismatic colours, more especially the .

blue and yellow, are useful guides as to the presence or absence of ozone in the air. If the air is very pure, the blue will be clear, and the F line will stand out distinctly; the yellow will show itself a true yellow, or have an orange tinge. On the other hand, when the reverse is the case, the yellow will have a greenish hue, and the blue will be deep and dark, with an indistinctness of the hydrogen line F. In most cases this peculiarity in the colours will be found to precede the atmospheric conditions.

I should recommend the observer to make it his first care, when using the spectroscope, to take note of any alteration in the true colours, as they will be found to change after looking for a short time through the instrument. The next thing is the amount of rainband (on sky free of cloud, if possible), at about 13 degrees all around, and then at the zenith. After this, begin at the blue end and work towards the red, carefully scanning the appearance of the constant lines, moist and dry air bands; notice if they are clear, indistinct, hazy, well-defined, or prominent, and finish up by noting the depth of shade at the red end— for a change of weather in certain conditions of the atmosphere is frequently heralded by its becoming clearer.

For the satisfactory prediction of weather it is highly necessary to keep a record of observations on a chart and in a note-book. On the former ought to be dotted down the percentage of rainband for eight points of the compass and zenith ; also that in the direction of the wind, or towards the quarter it is most likely to blow from in the course of the next twelve hours. The reason of the last is obvious, for we necessarily must expect the weather to come from that direction. In the note-book enter the peculiarities of the spectrum already noticed, and finally calculate, from what you have observed, the probabilities of rain or fine.

After a few months' practice, it is astonishing with what rapidity one can conduct a systematic observation. I use four spectroscopes of different dispersion in my atmospheric studies. A large two-prism one, with photo-

Fig. 5.

graphed micrometer (vide Fig. 5), arranged on a wooden alt-azimuth stand, with mirror for viewing the sun at different altitudes. It has three powers: the highest enlarging the rainband enormously, and enabling you to just discern two or three lines between the two D lines. A micro-spectroscope of small dispersion (Fig. 6). The first form of rain-band spectroscope (Fig. 1), which, being accustomed to, I feel reluctant to disqualify

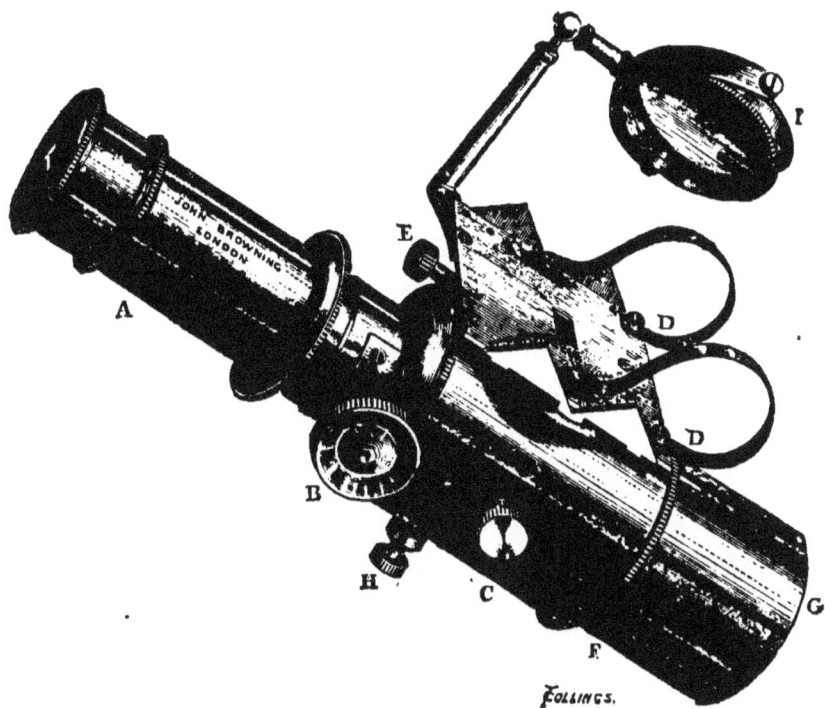

FIG. 6.

it for Grace's new and better form of hygro-spectroscope (Fig. 3). To the last I have adapted a stand (originally intended to hold a condenser), to steady it and meet the require-

3—2

ments of altitude and azimuth. Grace's is the
best all-round spectroscope, for reasons already
referred to in a former page, and taken in con-
junction with the other meteorological instruments will prove of great assistance to the
observer in prognosticating weather.

THE PORTABLE CLOUD MIRROR.

I have devised a very simple pocket cloud-
mirror to assist the observer in ascertaining
the direction from which clouds are travelling;
for it is a fact well known to meteorologists
that the direction the upper clouds (cirri) are
taking will often indicate the coming wind, two
or three days beforehand.   An outline of this
apparatus is given in Fig. 7.

It is constructed of a plain circular mirror,
two or three inches in diameter.   Let into
the centre is a small compass; from its cir-
cumference eight lines radiate to the outer
edge of the reflector; the termini of these lines

are equidistant from each other, and corre-
spond to the eight points of the compass.

To use the mirror it is only necessary to

The Portable Cloud Mirror

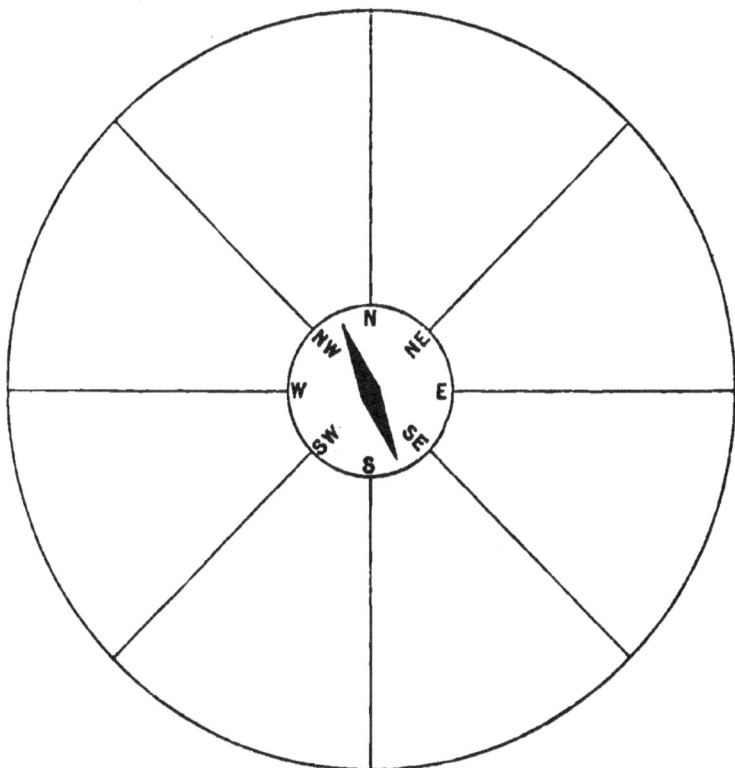

FIG. 7.

make one of the radiating lines coincide with
the direction of the needle, then, if requisite,
tilt the mirror, and observe the reflection of

some sharp edge of cloud, and by the help of the lines notice from which direction it is travelling. This can frequently be accomplished in a few seconds.

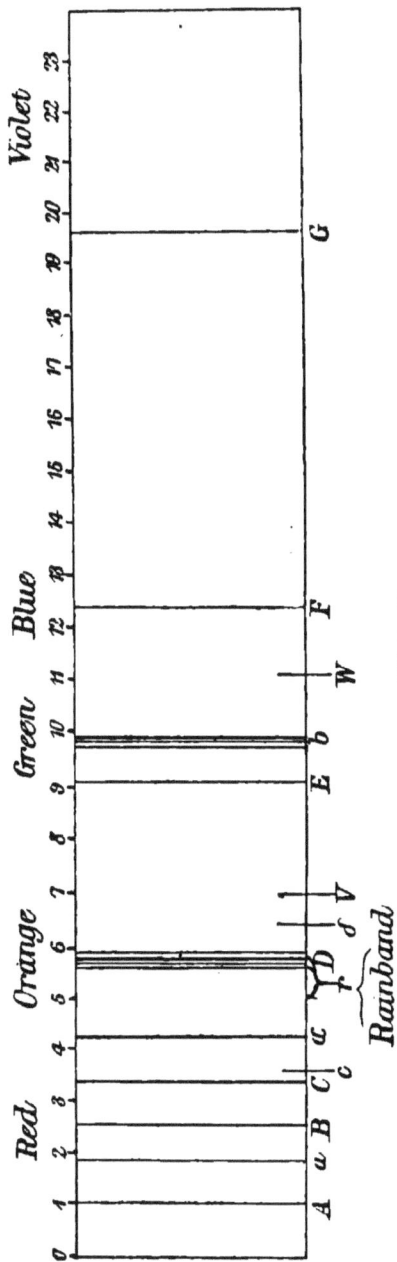

Plate I.

EXPLANATION OF THE PLATES.

Diagrams of: Percentage of cases of rain and snow for different intensities of rainband—Position of lines in spectrum—Lines forming the rainband.

PLATE I.

In this chart the actual appearances of the lines and bands of the spectrum are not intended to be represented, but merely their relative positions as seen in Grace's spectroscope. In the case of the bands, such as δ and v, the line is drawn as nearly as possible in the centre of the shading.

To the red side of D will be noticed the two lines, or rather, under sufficient amplification, two sets of lines, of the principal rainband.

Plate II.

Here are shown the vapour lines comprising the rainband. The two sets of lines before mentioned are delineated at *a* and *b*. The heavy, moderate, and slight lines are correctly placed and proportioned, and the very fine lines only approximately so. Their thickness varies with different conditions of atmospheric moisture. In the construction of the diagram a two-prism spectroscope with photographed micrometer was used, the light of the sun at a low altitude falling direct on the slit of the instrument.

Plate III.

The reader will see at a glance, in this Plate, the percentage of cases of rain and snow for each proportionate amount of rainband. The number of times that rain occurs within twenty-four hours increases with the darkness of the rainband, and *vice versâ*. On the other hand, the rainband will be

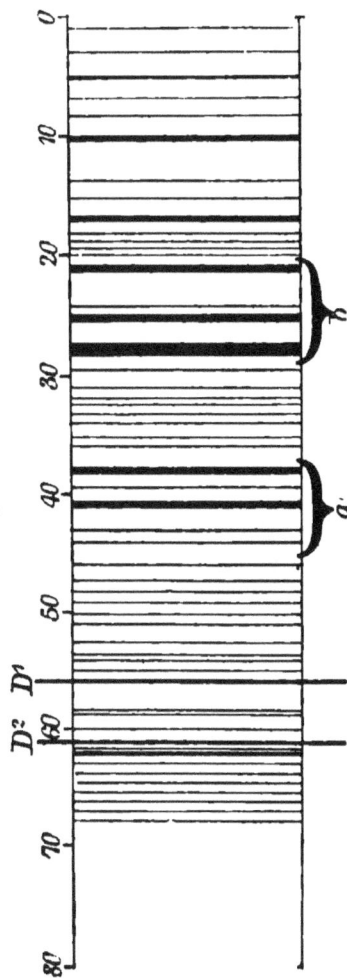

Lines in the Spectrum comprising the Principal Rainband

Plate II.

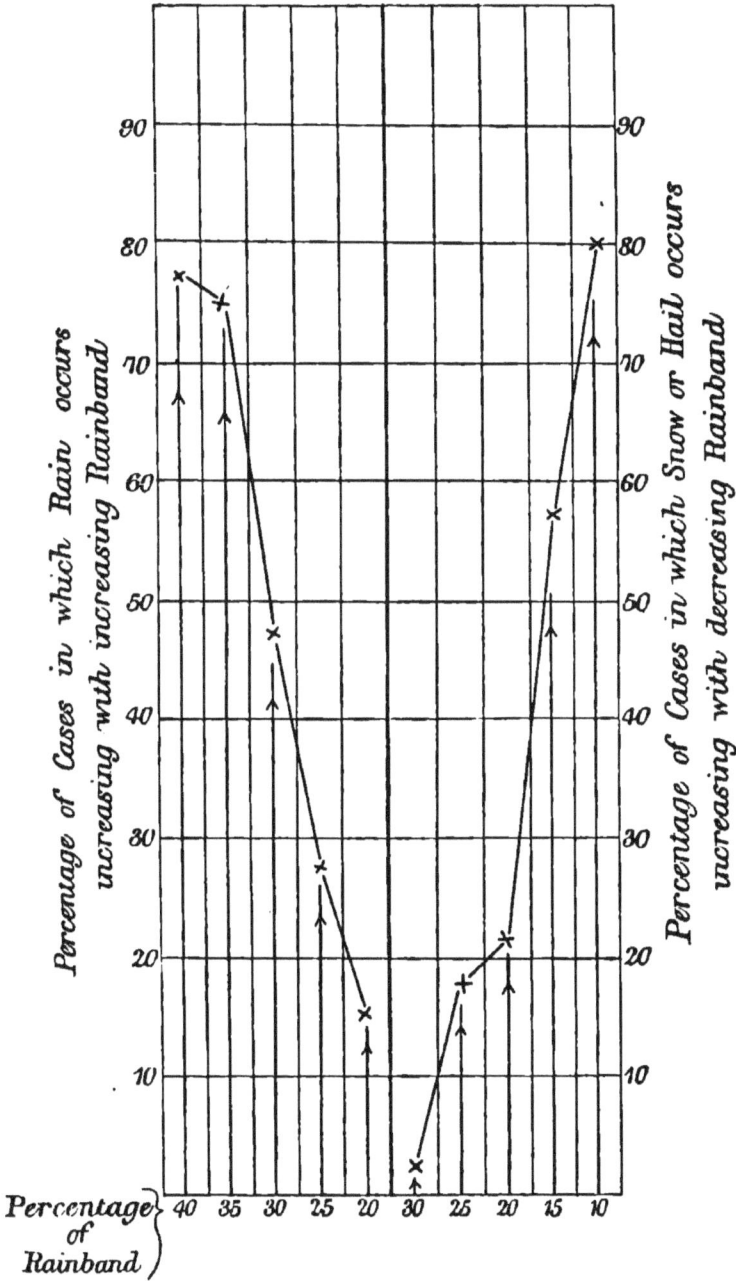

1883

Rain · Snow or Hail

Percentage of Cases in which Rain occurs increasing with increasing Rainband

Percentage of Cases in which Snow or Hail occurs increasing with decreasing Rainband

Plate III.

Percentage of Rainband} 40 35 30 25 20 · 30 25 20 15 10

observed to decrease with an increase in the number of instances in which snow falls.

There are two apparent discrepancies which demand a little explanation. I refer, firstly, to the tendency of the line on the rain side to rise at 20 per cent.; this is due to the fact that five cases of cold rain were included in the computation of the percentage ; but I think it may be almost safely surmised that these five ought to have been made use of on the snow-side of the diagram, as it is highly probable that they were melted snow; if this be admitted, and the alteration made, the lines connecting the percentage will present a much more direct course. Secondly, the number of cases at 40 per cent. of rainband one would naturally expect to be more, and probably rather less at 35 per cent., than is represented. It is possible that errors in calculating amount of rainband may account for these anomalies.

The conclusions to be gathered from this diagram are the following : That where the rainband is completely absent, snow will almost invariably occur. In all cases of 45 per cent. and upwards, rain will fall within twenty-four hours.

These lead one to the indisputable inference that the rainband increases before rain, and diminishes before snow.

APPENDIX.

THE following correspondence and leading article appeared in the *Times* during September, 1882. Since then the author has had, with only one exception, no reason to modify his experience as detailed in the two letters that he contributed. In the second rule of his first communication he advises, in a general way, to point the spectroscope in a northerly direction, according to the plan adopted by Professor Piazzi Smyth; but from further experience, and greater familiarity with the instrument, better results were obtained by observing towards the direction of the wind, or to the point it was most likely to change to in the course of the day. In 1882 the deductions from the hygro-spectroscope were somewhat wrapped in mystery—there

4

arc still many points that require clearing up ;
but the author feels sure that the reader, by
carefully attending to the rules laid down in
the preceding account, will meet with success,
and experience an increasing interest in the
indications of an instrument of such telescopic
power.

———————

From the *Times* of Sept. 12th, 1882.

A correspondent writes under date Edinburgh, Sept. 8th :

'In this uncertain climate of ours—
" variable as the shade "—everything that
bears on the forecasting of the weather is of
interest and importance. We have not heard
much hitherto of the spectroscope as a gauge
of the atmosphere. In Scotland we have had
this week what appears to be a very striking
instance of its trustworthiness in that capacity.
On Tuesday morning the following letter
appeared in the *Scotsman :*

'"Sir,

'"Last Friday morning the spectroscopic 'rainband' was the blackest and most intense of the season ;
and your issue of Saturday morning announced destruc-

tive floods, from most heavy rainfalls in various parts of Scotland, to have occurred on that day—Friday.

' " But this morning—Monday, September 4th—there is an absence of the 'rainband,' and a clearing away of all the watery vapour lines in the spectrum of sky-light, to an extent not equalled during the last two or three months.

' " In a powerful spectroscope the two solar D lines now stand out clear and clean, in place of being almost lost, as all through last month, in a thicket of terrestrial water-vapour lines. So the farmers may be enabled to gather in their crops at last, dry and in good condition, though, probably, in rather cold and sharp weather.

' " I am, etc.,

' " C. P. S."

' The writer is quite well known to be Mr. Piazzi Smyth, Astronomer Royal for Scotland, and Professor of Astronomy in the University of Edinburgh. How far, then, has his prediction been fulfilled ? The answer is—amply up to this date. Since Sunday we have had a succession of magnificent harvest days. Tuesday morning was rather dull, and it appeared at first as if the very confident forecast were about to be belied at once ; but the clouds cleared off in the forenoon, and there was brilliant sunshine with a cloudless sky

during the greater part of the day. Wednesday
was a repetition of Tuesday; and as the wind
blew from the north-west, the air was "rather
cold and sharp," as the Professor's letter
predicted. What made the case all the more
curious and striking, was the fact that on both
these days the forecasts of the Meteorological
Office told us to expect "showery and unsettled" weather over the greater part of
Scotland. The prediction for Wednesday was
"fair at first, then wet and unsettled." It
was more than "fair," not only "at first,"
but all day it was unusually bright and clear.
During Tuesday and Wednesday the barometer
rose steadily till it reached 30·388. It fell a
little on Thursday, but it stood at 30·335 at
night. The forecast of the Meteorological
Office for Thursday was, "South-westerly to
north-westerly winds, increasing in force;
becoming unsettled and rainy." In point of
fact, the wind was westerly all day; and,
instead of increasing in force, it fell away
toward evening to a calm. The weather did
not become "unsettled and rainy;" not a drop

of rain fell all day. The evening seemed
more settled than the morning ; and at eleven
o'clock at night the stars sparkled brilliantly,
and there was not a cloud in the sky., To-day
(Friday) opened with an over-clouded sky, but
no rain has yet fallen, although the official
forecast warns us that the day will be " fair
at first, then unsettled and rainy." As the
barometer has again taken an upward turn, we
may hope that this prediction will be again
falsified.

'These results appear to me to be very
striking and noteworthy. There has been
during the week a very marked conflict of
opinion between the chemists and the meteorologists, and thus far, at least, the chemists
have carried the day. The practical lesson
seems to be that more attention ought to be
given than seems hitherto to have been the
practice to the spectroscopic analysis of the
atmosphere. There may have been exceptional circumstances in the atmospheric
conditions of the week of which I am not
aware, and which have favoured the use of

the spectroscope. But the indubitable fact remains that in this instance the spectroscopic prevision has been amply justified and proved to be correct, while the ordinary meteorological forecasts have been from day to day entirely wrong.'

From the *Times* of Sept. 14th, 1882.

SIR,

As the subject of forecasting rain by means of the amount of rainband in the spectroscope has been brought forward by one of your correspondents, I beg to add my quota of experience in regard to it. For the sake of brevity, I will merely give a *résumé* of the result of my daily examinations with this instrument.

1. In taking observations it is very important to have both the slit in the spectroscope and the focus properly adjusted.

2. In a general way it is best to confine the observations to one particular part of the sky in a northerly direction, and at an angle of between 10° and 20° from the horizon.

3. If the amount of rainband is 20 per cent. or under, there will be fine weather— certainly for about six hours.

4. If the percentage of rainband, say at 7·a.m., is 60 and at 10 a.m. it has decreased to 30, that is no reason that fine weather may follow for that day, although probably no heavy rain will take place.

5. If the spectroscope be directed to the zenith, and 80 per cent. of the dark band is shown, a downpour of rain will certainly happen before long. I have observed this several times with a high and steady barometer, and at the time not much appearance of rain —notably on September 5, when the spectroscope indicated rain, but the other instruments did not.

From a careful trial of this valuable adjunct to the study of weather, I have come to the conclusion that the use of it, meteorologically speaking, has been most strangely neglected ; for I feel certain that our forecasts would be considerably improved if the observers at the various stations in connection with the

Meteorological Office were to report at the same time with their other observations of wind, temperature, etc., the percentage of rainband in the spectroscope. The price of it places it within reach of most persons ; and, as regards convenience of portableness, it is easily carried in the waistcoat pocket. Of course, a little practice is required to read the percentage of rainband, but I have found some take to it quite naturally.

I am, sir, yours obediently,

F. W. CORY, F.M.S.

Buckhurst-hill, Essex, Sept. 12th.

From the *Times* of Sept. 19th, 1882.

SIR,

The letters which have recently appeared in the *Times* under this heading testify to so much interest in the subject, that I venture to send the following remarks, which are the result of prolonged observation on every side of the question.

But first let me explain to your general readers what the spectroscopic ' rainband ' is.

If we look through a spectroscope directed to
any portion of the sky, we see a spectrum
tinted riband, crossed with thin black lines.
One of the strongest of these, situate in the
orange, is technically known as the D line.
When the instrument is directed to a pure
blue sky, this line is thin and faint; but in
certain conditions of cloud or sky, the red side
of the line has a dark shadow, as if it had
been shaded down with a bit of rough black
chalk. This shading is the rainband of
Professor P. Smyth, who maintains that the
relative darkness of- this band is a measure of
the nearness or quantity of rain.

Unfortunately, my own observations show
that, though there is scarcely any exception
to the appearance of the rainband being
speedily followed by rain, yet there are
numberless cases of impending or actual rain
during which no band is visible : and that, at
the same moment, one portion of the sky will
give a band, while a neighbouring portion will
show none at all. With one exception I
have never seen a rainband except when

common weather lore would have said it was going to rain, while in many cases I have observed such trustworthy prognostics as haloes when no rainband was visible.

In fact, the rainband appears to be simply a new sky prognostic, in many respects inferior to those in current use, though in a few cases affording information which they do not.

Of what use is any sky prognostic in weather forecasting? The whole system of modern meteorology turns round the shape of isobaric lines as .seen on charts similar to those published daily in the *Times*. Observation has shown that every shape of isobarics has a characteristic weather and appearance, so that there is little use in telegraphing up the occurrence of the rainband; for, given the chart, a meteorologist can write down on it not only the general weather, but the position of most of the best-known prognostics at the moment. His forecasts are based on the estimate he forms of the change which is likely to take place in the shape of the isobars during the next twenty-four hours. The

nature of these changes is still very imperfectly
known, but they certainly do not admit of any
mathematical calculation, like the situation of
a planet. The position of a forecaster is more
like that of a physician, who, although he has
classified the symptoms and ordinary course of
any disease, is still obliged to rely on his own
judgment, to a great extent, in giving the
prognosis of each particular case.

These are the principles on which the forecasts issued by the Meteorological Office are
based. For the days, and at the station from
which your Scotch correspondent writes, the
forecasts were certainly not so successful as
could be wished for, or as is usually the
case. The causes of failure cannot, of course,
be discussed in the columns of the *Times ;* but
I doubt whether the public have any idea of
the difficulty of forecasting, and even of check-
ing broken or unsettled weather like that to
which he alludes. Under those conditions,
rainfall is so local that half an inch may fall
at one place, while twenty miles off not a drop
will be seen ; and from the description he

gives of his weather, I strongly suspect that, if more records were obtained from the district to which the forecasts applied, they would be found to be more successful than he is inclined to think.

Anyhow, it is certain that a forecaster who relied on the spectroscope only would meet with most disheartening failures; and though there is doubtless a rich field of research open to the student of spectral vapour lines, I fear that meteorologists have little to hope for from the spectroscope in forecasting weather.

Yours obediently,

RALPH ABERCROMBY, F.M.S.

21, Chapel-street, S.W., Sept. 16th.

From the *Times* of Sept. 21st, 1882.

SIR,

Mr. R. Abercromby's ' prolonged observation ' of the spectroscope and weather forecasting must have been made with a very inefficient instrument, and with a very limited knowledge of the objects of observation; otherwise it would be difficult to understand

why he should describe the so-called rainband at D simply as a 'shading,' without
reference to the very pronounced lines seen in
it when there is much moisture in the air; or
why he should state that Professor P. Smyth
'maintains that the relative darkness of this
band is a measure of the nearness or quantity
of rain.'

The Astronomer Royal for Scotland is well
able to take care of himself; and I would not
advert to this travesty of his teaching but
that some of those who are proposing to use
the spectroscope for meteorological purposes
may be misled.

Aqueous vapour absorbs light at several
parts of the spectrum, but principally—or
perhaps I should rather say more visibly—on
the red side of D, and near—on the blue side
of—c. These absorptions are simply a measure
of the amount of aqueous vapour in the atmosphere. (For reasons which I will not enter
upon now, I believe that the absorption at the
last-named portion of the spectrum, termed by
Professor P. Smyth ' o,' is due to a grosser

molecule of aqueous vapour than that which absorbs near D. With a thick mist there is often an almost entire absence of absorption near D, while at c the absorption is strong. c and the band at D do not vary together.) Temperature must therefore be considered before the darkness of the rainband can be interpreted as an indication of rain.

When we remember the enormously greater thickness of the earth's atmosphere that a ray of sunlight has to traverse when we observe on the horizon than when we observe near the zenith ; when we remember, further, that the wind may have just commenced to bring up quantities of aqueous vapour from a particular quarter, it will be evident that Mr. Abercromby's statement ' that at the same moment one portion of the sky will give a band, while a neighbouring portion will show none,' may be quite correct without affecting the question at issue.

I fear to trespass further on your space, but I trust I have said enough to indicate the true answers to Mr. Abercromby's difficulties.

I will only point out, in conclusion, that as a knowledge of the isobars can only tell us the probable direction of the wind, a knowledge of the amount of moisture the wind is bearing —more accurate because more general than that afforded by the hygrometer—must be an important factor in weather forecasting.

I am, sir, yours obediently,

J. F. D. D.

Sept. 19th.

From the *Times* of Sept. 22nd, 1882.

SIR,

I can confirm all the warnings of your correspondent, Mr. Ralph Abercromby, on the difficulty and uncertainty attending spectroscopic forecasting of weather.

I have lately concluded a yachting cruise of seven weeks on the west coast of Scotland. I had with me a pocket spectroscope of admirable clearness of definition, made by Mr. Browning, of London; and I observed it frequently during all varieties of weather. I never could see that the 'rainband' had any predictive value. It was present frequently

in weather comparatively settled, and it was
sometimes evanescent or absent during fine
intervals of weather which was very wet and
very broken.

I must add, however, that the ordinary
aneroid barometer was equally useless for
predictive purposes. It was unusually high
for days together, during which the weather
was very unsettled, with violent rain, and
even some severe gales. It completely failed
to indicate one gale of exceptional violence,
which happened on the night of August 20 ;
and generally I may say that throughout the
season it has risen and fallen contempo-
raneously with the changes of weather, and
not in anticipation of them.

One very remarkable fact, however, I have
observed this season, as I have often observed
it before—that our barometer here almost invariably indicates the raging of great gales
over the western and even the southern shores
of England.

Mr. Piazzi Smyth certainly made a ' good
shot ' in his recent letter to the *Scotsman*, pre-

dicting some fine dry weather from a spectroscopic forecast. And it may be that the laws regulating the appearances of the 'rainband' are yet capable of yielding good results to skilled observers. But I suspect the skill is not very easily attainable with our present knowledge. Your obedient servant,

ARGYLL.

Inverary, Sept. 20th.

From the *Times* of Sept. 23rd, 1882.

SIR,

What may be done with the spectroscope in the matter of weather is, for the present at least, confined almost entirely to the question of rain—as, Will it rain, or will it not ; and, if it will, heavily or lightly ? The manner in which the spectroscope accomplishes this useful part is by its capacity for showing whether there is more or less than the usual quantity of watery vapour permeating the otherwise dry gases in the upper parts of the atmosphere, this watery vapour not being by any means the visible clouds themselves, but the invisible water-gas out of which they have to be formed, and by means of

5

which, when over-abundant, they obtain their privilege for enacting rainfall. So that never were wiser words uttered and more terse philosophy than those which are to be found in the ancient Book of Job, wherein, of the wondrously ' balanced clouds ' high up in mid-air, it is said, 'They pour down rain according to the vapour thereof.'

More or less of this water-vapour is always in the air, even on the very clearest days, and a happy thing for men that it is so ; for, as Dr. Tyndall and others have well shown, it moderates the excesses of hot solar radiation by day and cold radiation of the sky at night, and is more abundant in the hotter than the colder parts of the earth. Wherefore, according largely to its temperature for the time being, the air—otherwise consisting almost entirely of nitrogen and oxygen—can sustain, and does assimilate, as it were, a specified amount of this watery vapour, invisibly to the naked eye, the microscope, or the telescope ; but not so to the instrument of recent times, the spectroscope. And if the air vertically above any one place becomes presently charged

with more than its usual dose of such trans-
parent watery vapour (as it easily may, by
various modes and processes of nature), the
spectroscope shows that fact immediately, even
while the sky is still blue; clouds soon after
form, or thicken if already formed, and rain
presently begins to descend.

But how does the spectroscope show to the
eye what is declared to be invisible in all
ordinary optical instruments ? It is partly
by its power of discriminating the differently
coloured rays of which white light is made up,
and partly by the quality impressed on the
molecules of water at their primeval creation,
but only recently discovered, of stopping out
certain of those rays so discriminated and
placed in a rainbow-coloured order by the prism
and slit of the spectroscope, but transmitting
others freely. Hence it is that on looking at the
light of the sky through any properly-adjusted
spectroscope we see, besides the Newtonian
series of colours from red to violet, and besides
all the thin, dark Fraunhofer, or solar origin-
ated lines, of which it is not my object now

to speak, we see, I say, in one very definite part—viz., between the orange and yellow of that row of colours, or 'spectrum,' as it is called—a dark, hazy band stretching across it. That is the chief band of watery vapour ; and to see it very dark, even black, do not look at a dark part of the sky or of black clouds therein, but look, rather, where the sky is brightest, fullest of light to the naked eye, and where you can see through the greatest length of such well-illumined air as at a low, rather than high, angle of altitude, and either in warm weather, or, above all, just before a heavy rainfall, when there is, and must be, an extra supply of watery vapour in the atmosphere. Any extreme darkness, therefore, seen in that water-vapour band beyond what is usual for the season of the year and the latitude of the place, is an indication of rain-material accumulating abnormally; while, on the other hand, any notable deficiency in the darkness of it, other circumstances being the same, gives probability of dry weather, or absence of rain for very want of material to make it ; and the band, has, therefore, been

called, shortly, ' the rainband.' Thus, also,
' rainband spectroscopes' have been specially
constructed by several most expert opticians
in size so small as to be carriable in the waistcoat pocket, but so powerful and true that a
glance of two seconds' duration through one
of them suffices to tell an experienced observer
the general condition of the whole atmosphere.
Especially, too, of the upper parts of it, where
any changes—as they take place there almost
invariably earlier than below—enable such an
observer to favour his friends around him with
a prevision of what they are likely soon to
experience.

As an example of what may be done, and
done easily, after a certain amount of experience and understanding of the subject has been
acquired, I append, from a lady's meteorological journal, her notes of the mean temperature of the air and the intensity of the rainband for each of the first fifteen days of the
present month; and in a final column have
entered the amount of rainfall measured at the
Royal Observatory, Edinburgh, on each of
those days. The darker the rainband, the

larger is the figure set down for it; and it will
be seen pretty plainly, on running the eye down
that column and the next one, that with an
intensity of either 0 or 1 no rain follows,
or, we might also say, can follow; but with
an intensity of 2 rainfall begins, and with
3 it may be very heavy. All these rainband notes have been made with a spectroscope no larger than one's little finger, pur-
chased some six years ago and taken on many
voyages and travels since then:

Date, September, 1882.	Mean Temperature of the Air.	Rainband Intensity.	Depth of Rain measured in gauge at Royal Observatory, Edinburgh.
	Deg. Fahr.		Inch.
Friday, 1st ...	57·1	3	·044
Saturday, 2nd ...	59·2	2	·353
Sunday, 3rd ...	58·6	2	·015
Monday, 4th ...	54·4	0	0
Tuesday, 5th ...	55·7	1	0
Wednesday, 6th ...	55·2	0	0
Thursday, 7th ...	53·8	1	0
Friday, 8th ...	59·4	0	0
Saturday, 9th ...	54·0	1	0
Sunday, 10th ...	57·0	1	0
Monday, 11th ...	52·2	1	·040
Tuesday, 12th ...	48·6	0	0
Wednesday, 13th	52·8	1	0
Thursday, 14th ...	49·5	3	·062
Friday, 15th ...	56·2	2	·570

But if so much can be done by so small a
spectroscope, the question may be well asked
whether more still might not be accomplished
with a bigger and more powerful one, especially seeing that the dispersive powers of
both chemical and astronomical spectroscopes
have in late years been increased to a most
astonishing extent. The question is important, and somewhat new as well. I propose,
therefore, to devote the remainder of my space
to its answer, rather than to the practical rules
for using the smaller instruments, especially,
too, as they have been already introduced to
the public, both by my friend Mr. Rand
Capron, in his pamphlet, 'A Plea for the
Rainband,' and by myself in the fourteenth
volume of the 'Edinburgh Astronomical Observations;' also in the journal of the Scottish
Meteorological Society, and in the September
number of the 'Astronomical Register' for
1877.

The greater part of higher power spectroscopes are not suitable to rainband work, for
their fields are usually too dark. But having

recently built up for myself a large-sized variety of the instrument, possessing, perhaps, the greatest combination of power with transparency yet attained, and having it always mounted in an upper chamber, looking out at an altitude of about 5° over the north-western horizon (or most suitably for rainband work), I will try to describe shortly its action therein.

The classical ' rainband,' which in the little instrument is merely a very narrow fringe to an almost infinitely thin black line, is so magnified laterally in the larger instrument as to fill the whole breadth of the field.   The thin black line before spoken of is now not only split into two, but each of these are strong, thick, sharply defined lines, separated from each other by six or seven times the breadth of either.   These are the celebrated Solar D lines, D1 and D2, arising from the sodium metalloid burning or incandescent in the sun.   They are, therefore, perfectly uninfluenced by changes of the terrestrial atmosphere, hot or cold, wet or dry, and are, therefore, invaluable as references for degree

of visibility of the water-vapour lines and
bands which rise or fall in intensity precisely
with those changes. There are several of
these earthly water-vapour lines and bands
in and between and about the D lines
themselves; then a long breadth of band
towards the red side of D1; then a pair of
lines not so widely apart as the D lines, but
sometimes just as sharp and black; then two
or three fainter bands; then a grand triple,
of which the nearer line sometimes attains
greater blackness than either D line; then
beyond that three distinct, equal-spaced,
isolated bands; and further away towards the
red a stretch of faint haze and hazebands.

All these go to make up the one thin rainband of the little spectroscopes; and I for-
tunately had, through the month of August
and the early days of September, occupied
myself each morning in noting the greater
or less intensity of each, and all these water-
vapour lines and bands in terms of the two
Solar constants D1 and D2; and every such
morning there was an abundance of details to

see, to recognise, and to measure. But on
the morning of Monday, September 4th, when
the little instrument had truly enough marked
0 on its very small scale, I almost started at
finding in the large instrument every member
of its long rainband group, unless it were a
vanishing trace of one or two of the strongest,
utterly gone; while the two D lines were in
their accustomed strength, but far greater
clearness, for now they were all alone in the
field, save the ultra thin Solar nickel line
between them and one or two others, equally
thin and Solar on their blue side. The stages
of perceptible shade of water-vapour lines
which had thus been swept away, between
their this day's invisibility and their tre-
mendous strength no longer before than the
previous Friday, might have been expressed
by a scale not divided into three parts only,
but into thirty; and implied such a very unusual amount of absence of water-vapour,
that I not only felt sure of no rain falling
either next day or perhaps for several days
after, but that the weather must also be

coming on colder as well. Therefore it was that I took the step of instantly writing as I did to a local paper, promising the perplexed farmers dry weather at last, though probably sharp and cold, to get in their crops.

And how was that expectation fulfilled? Various meteorologists in different parts of the country have already declared themselves well satisfied with it. But I would now beg further attention to the little daily register already quoted, showing that from and including that day, Monday, September 4th, up to and including the next Saturday, not a drop of rain fell at the Observatory. Between the following Sunday and Monday a drizzle, but only amounting to 0·04 inch, occurred, and after that there were three more days equally dry with the preceding ones. But on Thursday, the 14th, the rain-band re-appeared in both spectroscopes in all its force; rain began to fall the same day, and next day's measure at the Observatory amounted to more than half an inch. Wherefore it is to be hoped that the farmers had busied themselves effectively while the dry weather lasted, for the return of these spectral lines of watery vapour showed that their autumn opportunity was then gone by.

C. PIAZZI SMYTH,
Astronomer-Royal for Scotland.
15, Royal-terrace, Edinburgh, Sept. 20th.

From the *Times* of Sept. 25th, 1882.

SIR,

Allow me to trespass a little further on your valuable space, as I think the difficulties the Duke of Argyll and Mr. Ralph Abercromby have experienced can be easily explained; and, moreover, it seems to me that the arguments they advance against the spectroscope for meteorological purposes rather prove the value of it than otherwise—that is to say, if used on a large scale at numerous stations. I will simply mention the following facts to bear out my statement. For instance, in the centre of a cyclone it is generally fine, and a very slight rainband may be visible in the spectroscope. In such a case, rain is, comparatively speaking, far

off. Of course, its approach to the observer
depends mainly upon the rate of the pro-
gressive motion of the storm.

During wet weather the rainband is not
unfrequently almost absent. This, for the
most part, denotes finer weather to follow.
Again, in many instances I have observed a
strong rainband in beautiful weather with
light and innocent-looking cumuli floating in
the sky ; and, as I afterwards ascertained, rain
had fallen about the same time at a distance
of at least ten miles from this place. I have
good reason to believe from several experiments that, in whatever direction a strong
rainband is shown, in such a direction rain is
falling, or about to fall. The immense
advantage of this indication in forecasting
precipitation of aqueous vapour will be sufficiently obvious to most of your readers, so
that I need not dwell any further on this
point. I hope individual observers do not
infer from my letter in the *Times* of the 14th
that the spectroscope can be used alone in
forecasting. I only recommended it as a

valuable adjunct to the study of the weather.

I am, sir, yours obediently,

F. W. CORY, F.M.S.

Buckhurst-hill, Essex, Sept. 22nd.

From the *Times* of Sept. 23rd, 1882.

It is recorded of a former Minister of the United States in this country that he was asked, soon after the commencement of his sojourn among us, what he thought of the English climate. He replied that he had not experienced any climate, but plenty of weather. The abundance to which he thus bore testimony is unfortunately attended by the disadvantage of extreme irregularity; to the extent that much of the important business of life, and also many of its pleasures, are seriously hindered by the difficulty of making previous arrangements for the accomplishment of any project with which bad weather would interfere. From the gathering in of the harvest to the organization of a picnic-party, the weather either promotes or thwarts our schemes ; and hence there has never been any lack of watchfulness for conditions upon which prognostications concerning it might be safely founded. In nearly every village there is some patriarch who is popularly believed to be unusually gifted with weather wisdom, or to have paid unusual attention to slight signs which escape the unobserving, and who is consequently sought for and consulted alike by the frivolous and by the earnest. The actions of the lower animals have been studied for the sake of the powers of weather foresight which many of them are commonly believed to possess ; and much popular lore of this kind has been tersely summed up by Gilbert White, in the metrical version of forty rules for predicting rain which is appended to the 'Natural History of Selborne.' More than forty years since one Murphy undertook to predict the weather for each day a whole year in advance, and issued an almanack containing his predictions. Early in January one of these was fulfilled in an unexpected way, with the result that the shop of his publisher was literally besieged by an eager crowd of would-be purchasers, by the first of whom the existing edition was speedily exhausted, with the consequence that the disappointed remainder had to be kept in order and dispersed by guardians of the public peace. A new edition was quickly prepared; but the predictions for subsequent days were less lucky, and the almanack, although issued annually for a time, soon fell into total disrepute.

Other prophets of weather have since arisen, concerning some of whom the most charitable view would be to suspect them of insanity; but it is only within the last few years that scientific observations of meteorological changes have been conducted with sufficient precision to afford hope of trustworthy results. The power conferred by the telegraph of ascertaining from hour to hour what is occurring in remote places has been utilized for the purpose of making known the weather which is apparently on its way to these latitudes; and the predictions based upon this knowledge

have been fulfilled with sufficient frequency
to render it apparent that the method of
inquiry is a useful one, while, at the same
time, they have been falsified so often as to
show that some of the causes which may
interfere with the transit of weather are not
yet fully understood. A storm may appear
to be directly on its way to us, and yet,
before it reaches our shores, it may be so
deflected as never actually to arrive upon
them. Until the laws of atmospheric movement have been more fully investigated,
sources of error will not be entirely set aside;
but the existence of these does not affect the
truth that the forecasts now made are often of
great practical value to the country.

Our columns have lately contained a correspondence on a new aid towards predicting weather which is manifestly worthy of
attention. On Tuesday, the 5th of September,
a letter, since acknowledged to have been
written by Mr. Piazzi Smyth, Astronomer
Royal for Scotland, appeared in the *Scotsman*
newspaper ; and in this letter the writer pre-

dicted some days of fine weather, upon the
ground of the disappearance ef the usual spectroscopic evidence of the presence of watery
vapour in the atmosphere. His prediction,
although it was opposed to the forecasts of
the Meteorological Office, was justified by the
event in a striking manner; and it has since
given rise to an animated controversy. As
a contribution to this controversy, we print
to-day a letter from Mr. Smyth, in which he
enters somewhat fully into the *rationale* of his
observations; and this may be summarized in
the following manner. The light which comes
to us from the sun is, in fact, wave movement in an infinitely subtle fluid; and the
waves are of several different rapidities of recurrence, these differences impressing the eyes
as differences of colour. When light is
passed through a prism, the waves of the
most rapid recurrence are most refracted, or
bent out of their original course; while those
of the least rapid recurrence are least bent;
the former producing to the eye the sensation
of violet, the latter the sensation of red. In-

termediate between them are other colours, depending upon waves of intermediate rapidities, the general result being that the beam of light acted upon is spread out into a party-coloured band or spectrum. The light proceeding from the sun would furnish a continuous spectrum ; but in passing through the solar envelopes, or through the atmosphere, some portion of it is quenched by collision with wave movement precisely of the same pitch, the waves neutralizing each other, and producing stillness, when the summits of one series fall precisely into the hollows of the other. Hence the solar spectrum is traversed by a number of dark lines in definite positions, and one of these, which appears single in small instruments, although in larger ones it admits of separation into constituent parts, is due to the presence of watery vapour in the atmosphere. Mr. Smyth points out that this watery vapour is, so to speak, the raw material of rain, and that without it rain cannot be produced. He therefore infers that its total absence involves of necessity a

6—2

period of dry weather, and that its marked presence indicates a state of things in which the material for rain is abundant, and in which rainfall must, therefore, be regarded as highly probable.

Assuming the facts to be as stated, the letters of the correspondents who have expressed doubts of the utility of the spectroscope as a means of predicting weather do not appear to possess much value. No scientifically instructed person would expect the degree of abundance of watery vapour to be alone conclusive with regard to a result which no doubt depends partly upon this, and partly upon the concurrence of many variable factors. Watery vapour may be absent at a given moment, or in a given direction, and may be rapidly brought from elsewhere, just as in all probability it may sometimes be rapidly conveyed away, by atmospheric movement, without occasioning rainfall in a place where it was actually detected.

The condition first described by Mr. Smyth, in which the spectroscopic vapour-band had

absolutely disappeared, is not likely to be of
frequent occurrence ; and it would probably
be only something as marked as this which
would justify a conclusion based upon one
element of the question. The real value of
the spectroscope to meteorologists has yet to
be ascertained, and must depend upon the
power which it promises to afford of determining the amount or proportion of watery vapour
in the atmosphere at a given time, and possibly
also the molecular state of this vapour, and of
adding the facts thus made known to any
others which may be available for the purposes of the inquiry.

Temperature must always constitute an important element in predictions founded upon
the presence of watery vapour, because an
ordinary immediate cause of rain is the arrival
of moisture-laden air in a region colder than
that from which it came. It is not likely
that much will ever be learned from isolated
observations, taken with small instruments,
by which the precise composition of the
vapour-band is not disclosed, or taken with-

out reference to conditions which may co-exist in other localities.

Enough has at least been made out to show that the observation of the vapour-band is one which cannot be neglected for the future ; and it will be time to decide upon the precise import of the appearances which this band may present when they have been studied over extended areas, and with proper reference to all associated phenomena. The progress lately made in forecasting weather is sufficient to justify great hope of further advances· in the same direction ; but it is not sufficient to permit the neglect of any method of inquiry which promises to render our knowledge of any factor in the production of rainfall at all more methodical or complete.

THE END.

BILLING AND SONS, PRINTERS, GUILDFORD.

THE NEW SELF-REGISTERING ANEROID.

This instrument is extremely simple and hardy in construction. It has neither spring nor chain, the motive power being obtained from Seven Vacuum Chambers ; these are attached by levers to the arm carrying the pen, which registers the height of the Barometer on the diagram, by means of a long aluminium arm carrying a pen. The drum revolves once in a week, motion being given to it by an eight-day clock movement, which is out of sight inside the drum.

The diagrams from this instrument are not only interesting, but much more valuable than those given by an ordinary Barometer ; for instance, if an observer after setting a Mercurial Barometer at 30 inches, at 8 o'clock in the evening, looks at the Barometer at 8 o'clock the following morning, and finds that it registers 29·7, he will conclude that the Barometer has fallen 3·10ths, and is probably falling at that time, but the Self-Registering Aneroid might show him that the Barometer had fallen 5·10ths in the night, and had risen 2·10ths since. When a storm of wind or rain is experienced, observers who consult the Barometer only at long intervals, will at times remark that the instrument gave no indication of the storm, yet with the Self-Registering Aneroid, this would have been found a mistake, for the Barometer had probably fallen and risen again in the interval ; and it is well known that a rapid rise after a fall, indicates a stronger gale than the fall itself.

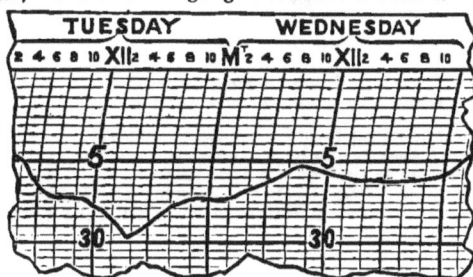

(*Small portion of a Weekly Diagram.*)

The price of the New Self-registering Aneroid Barometer with Fifty-two diagram papers is £6 0 0
Packing Box 0 1 6

JOHN BROWNING,

Optical and Physical Instrument Maker to Her Majesty's Government, the Royal Society, the Royal Observatories of Greenwich and Edinburgh, and the Observatories at Kew, Cambridge, etc., etc.,

63, STRAND, LONDON, W.C.

Established 100 years.

THE
RAINBAND SPECTROSCOPE.

For years the prediction of the weather has received constantly increasing attention; this makes the introduction of any new method of forecasting the weather of great importance. To nearly every person in the community a means of telling in the morning whether the day will be fine or wet is of considerable value, to farmers and many others whose work goes on exclusively in the open air the power of predicting this is of still greater consequence. The Barometer will frequently rise for hours and yet rain come on and continue falling. As a means of predicting coming rain no instrument has been introduced which equals the Rainband Spectroscope. *The results obtained by observers last year have proved the great value of the Rainband Spectroscope for predicting the coming of rain,* and the adoption of the instrument by Agriculturists might be a matter of national importance.

LIST OF PRICES.

	£	s.	d.
Rainband Spectroscope with fixed slit	1	12	6
Leather Case extra	0	1	0
Rainband Spectroscope of superior construction with adjustable slit in case	2	10	0
Grace's New Rainband Spectroscope with adjustable slit and fine focussing adjustment in morocco case complete	3	8	6

This instrument is of an improved optical construction, and it shows the rainband as *separate lines.* It is by far the easiest instrument to use for those who are unaccustomed to the use of a Spectroscope.

A coloured diagram of the Spectrum, showing the Rainband of various degrees of intensity, is given in "How to Work with the Spectroscope," by John Browning, price 1s. 6d.

JOHN BROWNING,

Optical and Physical Instrument Maker to Her Majesty's Government,

63, STRAND, LONDON, W.C.

Established 100 years.

JOHN BROWNING'S
PRICES OF SPECTACLES.

(Including suiting the Sight by Correspondence, respecting
which Particulars will be sent free.)

	Per pair
Good steel Spectacles, with glass lenses	4/6
Superior light steel frames, with best glass lenses	7/6
Superior light steel frames, with Brazilian pebble lenses	10/6
Very superior light steel frames, with best Brazilian pebble lenses ...	15/-
Best light steel frames, with best Axis cut pebble lenses	21/-
Invisible steel Spectacles, with hook or curled sides and grooved lenses for the frame to fit into, thereby giving them a very light appearance	10/6
Best Invisible steel Spectacles, as above described	15/-

Gold Spectacles, from 18/6 to 70/- per pair, according to weight and
quality.

Good steel Spectacles, with tinted lenses perfectly plane for protecting the eyes against strong light	4/6
Superior light steel frames, with coloured plane glass as above ...	7/6
Best steel frames, with coloured plane glass	10/6
Superior light steel frames, with Concave or Convex lenses of coloured glass for protecting the eyes against strong light	10/6
Best steel frames, with coloured lenses, as above	15/-
Best tempered steel Spectacles with globular glasses, smoke coloured or blue tinted	10/6
Best tempered steel Spectacles, with glasses as above and wire or silk gauze sides for use in India or Egypt	15/-
Best steel Spectacles D shaped eyes with glass sides	10/-
The New Patent Preservers attached to Spectacle Frames for protecting the eyes from top light, particularly gas, complete with lenses, price from	5/6

Cases in every instance are included in the price.

PRICES OF SPRING FOLDERS.

(Including suiting the Sight by Correspondence.)

Good steel frames, with clear glass lenses	4/6
Superior light steel frames, with best glass lenses	7/6
Best light straw steel frames, with best grooved lenses	10/6
Best light steel frames, nickelized to prevent rust, and fitted with best glass lenses	15/6

Gold folders from 18/6 to 70/- per pair, according to weight and quality.
Best Brazilian pebbles fitted to any of the above, 5/- extra.
Browning's Axis cut pebbles, 10/- extra.

Cases in every instance are included in the price.

JOHN BROWNING,
*Optical and Physical Instrument Maker to Her Majesty's
Government.*
63, STRAND, LONDON, W.C.
_ **Established 100 years.**

CHATTO & WINDUS'S

LIST OF BOOKS.

* * * * * * * * * * * * * *

About.—The Fellah: An Egyptian Novel. By EDMOND ABOUT. Translated by Sir RANDAL ROBERTS. Post 8vo, illustrated boards, **2s.** ; cloth limp, **2s. 6d.**

Adams (W. Davenport), Works by:

A Dictionary of the Drama. Being a comprehensive Guide to the Plays, Playwrights, Players, and Playhouses of the United Kingdom and America, from the Earliest to the Present Times. Crown 8vo, half-bound, **12s. 6d.**

Latter-Day Lyrics. Edited by W. DAVENPORT ADAMS. Post 8vo, cloth limp, **2s. 6d.**

Quips and Quiddities. Selected by W. DAVENPORT ADAMS. Post 8vo, cloth limp, **2s. 6d.**

Advertising, A History of, from the Earliest Times. Illustrated by Anecdotes, Curious Specimens, and Notices of Successful Advertisers. By HENRY SAMPSON. Crown 8vo, with Coloured Frontispiece and Illustrations, cloth gilt, **7s. 6d.**

Agony Column (The) of "The Times," from 1800 to 1870. Edited, with an Introduction, by ALICE CLAY. Post 8vo, cloth limp, **2s. 6d.**

Aide (Hamilton), Works by:

Carr of Carrlyon. Post 8vo, illustrated boards, **2s.**

Confidences. Post 8vo, illustrated boards, **2s.**

Alexander (Mrs.).—Maid, Wife, or Widow? A Romance. By Mrs. ALEXANDER. Post 8vo, illustrated boards, **2s.** ; cr. 8vo, cloth extra, **3s. 6d.**

Allen (Grant), Works by:

Colin Clout's Calendar. Crown 8vo, cloth extra, **6s.**

The Evolutionist at Large. Crown 8vo, cloth extra, **6s.**

Vignettes from Nature. Crown 8vo, cloth extra, **6s.**

Architectural Styles, A Handbook of. Translated from the German of A. ROSENGARTEN, by W. COLLETT-SANDARS. Crown 8vo, cloth extra, with 639 Illustrations, **7s. 6d.**

Art (The) of Amusing: A Collection of Graceful Arts, Games, Tricks, Puzzles, and Charades. By FRANK BELLEW. With 300 Illustrations. Cr. 8vo, cloth extra, **4s. 6d.**

Artemus Ward:

Artemus Ward's Works: The Works of CHARLES FARRER BROWNE, better known as ARTEMUS WARD. With Portrait and Facsimile. Crown 8vo, cloth extra, **7s. 6d.**

Artemus Ward's Lecture on the Mormons. With 32 Illustrations. Edited, with Preface, by EDWARD P. HINGSTON. Crown 8vo, **6d.**

The Genial Showman: Life and Adventures of Artemus Ward. By EDWARD P. HINGSTON. With a Frontispiece. Crown 8vo, cloth extra, **3s. 6d.**

Ashton (John), Works by:

A History of the Chap-Books of the Eighteenth Century. With nearly 400 Illusts., engraved in facsimile of the originals. Cr. 8vo, cl. ex., 7s. 6d.

Social Life in the Reign of Queen Anne. From Original Sources. With nearly 100 Illusts. Cr.8vo,cl.ex.,7s.6d.

Humour, Wit, and Satire of the Seventeenth Century. With nearly 100 Illusts. Cr. 8vo, cl. extra, 7s. 6d.

English Caricature and Satire on Napoleon the First. With 120 Illustrations from the Originals. Two Vols., demy 8vo, 28s. [*In preparation.*

Bacteria.—A Synopsis of the

Bacteria and Yeast Fungi and Allied Species. By W. B. GROVE, B.A. With over 100 Illustrations. Cr. 8vo, cloth extra, 3s. 6d. [*In preparation.*

Balzac's "Comedie Humaine"

and its Author. With Translations by H. H. WALKER. Post 8vo, cl.limp,2s. 6d.

Bankers, A Handbook of Lon-

don; together with Lists of Bankers from 1677. By F. G. HILTON PRICE. Crown 8vo, cloth extra, 7s. 6d.

Bardsley (Rev. C.W.),Works by:

English Surnames: Their Sources and Significations. Cr.8vo,cl. extra, 7s.6d.

Curiosities of Puritan Nomenclature. Crown 8vo, cloth extra, 7s. 6d.

Bartholomew Fair, Memoirs

of. By HENRY MORLEY. With 100 Illusts. Crown 8vo, cloth extra, 7s. 6d.

Beauchamp. — Grantley

Grange: A Novel. By SHELSLEY BEAUCHAMP. Post 8vo, illust. bds., 2s.

Beautiful Pictures by British

Artists: A Gathering of Favourites from our Picture Galleries. In Two Series. All engraved on Steel in the highest style of Art. Edited, with Notices of the Artists, by SYDNEY ARMYTAGE, M.A. Imperial 4to, cloth extra, gilt and gilt edges, 21s. per Vol.

Bechstein. — As Pretty as

Seven, and other German Stories. Collected by LUDWIG BECHSTEIN. With Additional Tales by the Brothers GRIMM, and 100 Illusts. by RICHTER. Small 4to, green and gold, 6s. 6d.; gilt edges, 7s. 6d.

Beerbohm. — Wanderings in

Patagonia; or, Life among the Ostrich Hunters. By JULIUS BEERBOHM. With Illusts. Crown 8vo, cloth extra, 3s. 6d.

Belgravia for 1884. One

Shilling Monthly, Illustrated by P. MACNAB.—Two Serial Stories are now appearing in this Magazine: "The Lover's Creed," by Mrs. CASHEL HOEY; and "The Wearing of the Green," by the Author of "Love the Debt."

* Now ready, the Volume for NOVEMBER, 1883, to FEBRUARY, 1884, cloth extra, gilt edges, 7s. 6d.; Cases for binding Vols., 2s. each.

Belgravia Holiday Number.

With Stories by JAMES PAYN, F. W. ROBINSON, J. ARBUTHNOT WILSON, and others. Demy 8vo, with Illustrations, 1s. [*Preparing.*

Bennett(W.C.,LL.D.),Works by:

A Ballad History of England. Post 8vo, cloth limp, 2s.

Songs for Sailors. Post 8vo, cloth limp, 2s.

Besant (Walter) and James

Rice, Novels by. Post 8vo, illust. boards, 2s. each; cloth limp, 2s. 6d. each; or crown 8vo, cloth extra, 3s. 6d. each.

Ready-Money Mortiboy.
With Harp and Crown.
This Son of Vulcan.
My Little Girl.
The Case of Mr. Lucraft.
The Golden Butterfly.
By Celia's Arbour.
The Monks of Thelema.
'Twas in Trafalgar's Bay.
The Seamy Side.
The Ten Years' Tenant.
The Chaplain of the Fleet.

Besant (Walter), Novels by:

All Sorts and Conditions of Men: An Impossible Story. With Illustrations by FRED. BARNARD. Crown 8vo, cloth extra, 3s. 6d.; post 8vo, illustrated boards, 2s.

The Captains' Room, &c. With Frontispiece by E. J. WHEELER. Crown 8vo, cloth extra, 3s. 6d.; post 8vo, illustrated boards, 2s.

All in a Garden Fair. Three Vols., crown 8vo.

Dorothy Forster. Three Vols., crown 8vo. [*Shortly.*

Betham-Edwards (M.), Novels

by. Crown 8vo, cloth extra, 3s. 6d. each.: post 8vo, illust. bds., 2s. each.

Felicia. | Kitty.

Bewick (Thomas) & his Pupils.
By AUSTIN DOBSON. With 100 Illustrations. Square 8vo, cloth extra, 10s. 6d. [*Preparing.*

Birthday Books:—
The Starry Heavens: A Poetical Birthday Book. Square 8vo, handsomely bound in cloth, 2s. 6d.

Birthday Flowers: Their Language and Legends. By W. J. GORDON. Beautifully Illustrated in Colours by VIOLA BOUGHTON. In illuminated cover, crown 4to, 6s.

The Lowell Birthday Book. With Illusts., small 8vo, cloth extra, 4s. 6d.

Bishop.—Old Mexico and her
Lost Provinces. By WILLIAM HENRY BISHOP. With 120 Illustrations. Demy 8vo, cloth extra, 10s. 6d.

Blackburn's (Henry) Art Hand-
books. Demy 8vo, Illustrated, uniform in size for binding.

Academy Notes, separate years, from 1875 to 1883, each 1s.

Academy Notes, 1884. With Illustrations. 1s. [*Preparing.*

Academy Notes, 1875–79. Complete in One Vol.,with nearly 600 Illusts. in Facsimile. Demy 8vo, cloth limp, 6s.

Grosvenor Notes, 1877. 6d.

Grosvenor Notes, separate years, from 1878 to 1883, each 1s.

Grosvenor Notes, 1884. With Illustrations. 1s. [*Preparing.*

Grosvenor Notes, 1877–82. With upwards of 300 Illustrations. Demy 8vo, cloth limp, 6s.

Pictures at South Kensington. With 70 Illustrations. 1s.

The English Pictures at the National Gallery. 114 Illustrations. 1s.

The Old Masters at the National Gallery. 128 Illustrations. 1s. 6d.

A Complete Illustrated Catalogue to the National Gallery. With Notes by H. BLACKBURN, and 242 Illusts. Demy 8vo, cloth limp, 3s.

The Paris Salon, 1884. With over 300 Illusts. Edited by F. G. DUMAS. Demy 8vo, 3s. [*Preparing.*

The Art Annual, 1883–4. Edited by F. G. DUMAS. With 300 full-page Illustrations. Demy 8vo, 5s.

Boccaccio's Decameron ; or,
Ten Days' Entertainment. Translated into English, with an Introduction by THOMAS WRIGHT, F.S.A. With Portrait, and STOTHARD's beautiful Copperplates. Cr. 8vo, cloth extra, gilt, 7s. 6d.

Blake (William): Etchings from
his Works. By W. B. SCOTT. With descriptive Text. Folio, half-bound boards, India Proofs, 21s.

Bowers'(G.) Hunting Sketches:
Canters in Crampshire. Oblong 4to, half-bound boards, 21s.

Leaves from a Hunting Journal. Coloured in facsimile of the originals. Oblong 4to, half-bound, 21s.

Boyle (Frederick), Works by :
Camp Notes: Stories of Sport and Adventure in Asia, Africa, and America. Crown 8vo, cloth extra, 3s. 6d. ; post 8vo, illustrated bds., 2s.

Savage Life. Crown 8vo, cloth extra, 3s. 6d. ; post 8vo, illustrated bds., 2s.

Brand's Observations on Pop-
ular Antiquities, chiefly Illustrating the Origin of our Vulgar Customs, Ceremonies, and Superstitions. With the Additions of Sir HENRY ELLIS. Crown 8vo, cloth extra, gilt, with numerous Illustrations, 7s. 6d.

Bret Harte, Works by :
Bret Harte's Collected Works. Arranged and Revised by the Author. Complete in Five Vols., crown 8vo, cloth extra, 6s. each.
Vol. I. COMPLETE POETICAL AND DRAMATIC WORKS. With Steel Portrait, and Introduction by Author.
Vol. II. EARLIER PAPERS—LUCK OF ROARING CAMP, and other Sketches —BOHEMIAN PAPERS — SPANISH AND AMERICAN LEGENDS.
Vol. III. TALES OF THE ARGONAUTS —EASTERN SKETCHES.
Vol. IV. GABRIEL CONROY.
Vol. V. STORIES — CONDENSED NOVELS, &c.

The Select Works of Bret Harte, in Prose and Poetry. With Introductory Essay by J. M. BELLEW, Portrait of the Author, and 50 Illustrations. Crown 8vo, cloth extra, 7s. 6d.

Gabriel Conroy: A Novel. Post 8vo, illustrated boards, 2s.

An Heiress of Red Dog, and other Stories. Post 8vo, illustrated boards, 2s. ; cloth limp, 2s. 6d.

The Twins of Table Mountain. Fcap. 8vo, picture cover, 1s. ; crown 8vo, cloth extra, 3s. 6d.

Luck of Roaring Camp, and other Sketches. Post 8vo, illust. bds., 2s.

Jeff Briggs's Love Story. Fcap 8vo, picture cover, 1s. ; cloth extra, 2s. 6d.

Flip. Post 8vo, illustrated boards, 2s. ; cloth limp, 2s. 6d.

Californian Stories (including THE TWINS OF TABLE MOUNTAIN, JEFF BRIGGS'S LOVE STORY, &c.) Post 8vo, illustrated boards, 2s.

Brewer (Rev. Dr.), Works by :

The Reader's Handbook of Allusions, References, Plots, and Stories. Third Edition, revised throughout, with a New Appendix, containing a COMPLETE ENGLISH BIBLIOGRAPHY. Cr. 8vo, 1,400 pp., cloth extra, 7s. 6d.

A Dictionary of Miracles: Imitative, Realistic, and Dogmatic. Crown 8vo, cloth extra, 7s. 6d.   [*Immediately.*

Brewster (Sir David), Works by:

More Worlds than One: The Creed of the Philosopher and the Hope of the Christian. With Plates. Post 8vo, cloth extra, 4s. 6d.

The Martyrs of Science: Lives of GALILEO, TYCHO BRAHE, and KEP-LER. With Portraits. Post 8vo, cloth extra, 4s. 6d.

Letters on Natural Magic. A New Edition, with numerous Illustrations, and Chapters on the Being and Faculties of Man, and Additional Phenomena of Natural Magic, by J. A. SMITH. Post 8vo, cloth extra, 4s. 6d.

Brillat-Savarin.—Gastronomy

as a Fine Art. By BRILLAT-SAVARIN. Translated by R. E. ANDERSON, M.A. Post 8vo, cloth limp, 2s. 6d.

Browning.—The Pied Piper of

Hamelin. By ROBERT BROWNING. Illust. by GEORGE CARLINE. Large 4to, illuminated cover, 1s.
[*In preparation.*

Burnett (Mrs.), Novels by :

Surly Tim, and other Stories. Post 8vo, illustrated boards, 2s.

Kathleen Mavourneen. Fcap. 8vo, picture cover, 1s.

Lindsay's Luck. Fcap. 8vo, picture cover, 1s.

Pretty Polly Pemberton. Fcap. 8vo picture cover, 1s.

Burton (Captain), Works by :

To the Gold Coast for Gold: A Personal Narrative. By RICHARD F. BURTON and VERNEY LOVETT CAMERON. With Maps and Frontispiece. Two Vols., crown 8vo, cloth extra, 21s.

The Book of the Sword: Being a History of the Sword and its Use in all Countries, from the Earliest Times. By RICHARD F. BURTON. With over 400 Illustrations. Square 8vo, cloth extra, 32s.

Buchanan's (Robert) Works :

Ballads of Life, Love, and Humour. With a Frontispiece by ARTHUR HUGHES. Crown 8vo, cloth extra, 6s.

Selected Poems of Robert Buchanan. With Frontispiece by T. DALZIEL. Crown 8vo, cloth extra, 6s.

Undertones. Cr. 8vo, cloth extra, 6s.

London Poems. Crown 8vo, cloth extra, 6s.

The Book of Orm. Crown 8vo, cloth extra, 6s.

White Rose and Red: A Love Story. Crown 8vo, cloth extra, 6s.

Idylls and Legends of Inverburn. Crown 8vo, cloth extra, 6s.

St. Abe and his Seven Wives : A Tale of Salt Lake City. With a Frontispiece by A. B. HOUGHTON. Crown 8vo, cloth extra, 5s.

The Hebrid Isles: Wanderings in the Land of Lorne and the Outer Hebrides. With Frontispiece by W. SMALL. Crown 8vo, cloth extra, 6s.

A Poet's Sketch-Book: Selections from the Prose Writings of ROBERT BUCHANAN. Crown 8vo, cl. extra, 6s.

The Shadow of the Sword: A Romance. Crown 8vo, cloth extra, 3s. 6d. ; post 8vo, illust. boards, 2s.

A Child of Nature: A Romance. With a Frontispiece. Crown 8vo, cloth extra, 3s. 6d.; post 8vo, illust. bds., 2s.

God and the Man: A Romance. With Illustrations by FRED. BARNARD. Crown 8vo, cloth extra, 3s. 6d. ; post 8vo, illustrated boards, 2s.

The Martyrdom of Madeline: A Romance. With Frontispiece by A. W. COOPER. Cr. 8vo, cloth extra, 3s. 6d.; post 8vo, illustrated boards, 2s.

Love Me for Ever. With a Frontispiece by P. MACNAB. Crown 8vo, cloth extra, 3s. 6d.; post 8vo, illustrated boards, 2s.

Annan Water: A Romance. Three Vols., crown 8vo.

The New Abelard: A Romance. Three Vols., crown 8vo.

Foxglove Manor: A Novel. Three Vols., crown 8vo.   [*In preparation.*

Robert Buchanan's Complete Poetical Works. With Steel-Plate Portrait. Crown 8vo, cloth extra, 7s. 6d.
[*In the press.*

Burton (Robert) :

The Anatomy of Melancholy. A New Edition, complete, corrected and enriched by Translations of the Classical Extracts. Demy 8vo, cloth extra, 7s. 6d.

Melancholy Anatomised : Being an Abridgment, for popular use, of BURTON'S ANATOMY OF MELANCHOLY. Post 8vo, cloth limp, 2s. 6d.

Bunyan's Pilgrim's Progress.

Edited by Rev. T. SCOTT. With 17 Steel Plates by STOTHARD, engraved by GOODALL, and numerous Woodcuts. Crown 8vo, cloth extra, gilt, 7s. 6d.

Byron (Lord):

Byron's Letters and Journals. With Notices of his Life. By THOMAS MOORE. A Reprint of the Original Edition, newly revised, with Twelve full-page Plates. Crown 8vo, cloth extra, gilt, 7s. 6d.

Byron's Don Juan. Complete in One Vol., post 8vo, cloth limp, 2s.

Cameron (Commander) and

Captain Burton.—To the Gold Coast for Gold: A Personal Narrative. By RICHARD F. BURTON and VERNEY LOVETT CAMERON. With Frontispiece and Maps. Two Vols., crown 8vo, cloth extra, 21s.

Cameron (Mrs. H. Lovett),

Novels by:

Juliet's Guardian. Post 8vo, illustrated boards, 2s.; crown 8vo, cloth extra, 3s. 6d.

Deceivers Ever. Post 8vo, illustrated boards, 2s.; crown 8vo, cloth extra, 3s. 6d.

Campbell.—White and Black:

Travels in the United States. By Sir GEORGE CAMPBELL, M.P. Demy 8vo, cloth extra, 14s.

Carlyle (Thomas):

Thomas Carlyle: Letters and Recollections. By MONCURE D. CONWAY, M.A. Crown 8vo, cloth extra, with Illustrations, 6s.

On the Choice of Books. By THOMAS CARLYLE. With a Life of the Author by R. H. SHEPHERD. New and Revised Edition, post 8vo, cloth extra, Illustrated, 1s. 6d.

The Correspondence of Thomas Carlyle and Ralph Waldo Emerson, 1834 to 1872. Edited by CHARLES ELIOT NORTON. With Portraits. Two Vols., own 8vo, cloth extra, 24s.

Chapman's (George) Works:

Vol. I. contains the Plays complete, including the doubtful ones. Vol. II., the Poems and Minor Translations, with an Introductory Essay by ALGERNON CHARLES SWINBURNE. Vol. III., the Translations of the Iliad and Odyssey. Three Vols., crown 8vo, cloth extra, 18s.; or separately, 6s. each.

Chatto & Jackson.—A Treatise

on Wood Engraving, Historical and Practical. By WM. ANDREW CHATTO and JOHN JACKSON. With an Additional Chapter by HENRY G. BOHN; and 450 fine Illustrations. A Reprint of the last Revised Edition. Large 4to, half-bound, 28s.

Chaucer:

Chaucer for Children: A Golden Key. By Mrs. H. R. HAWEIS. With Eight Coloured Pictures and numerous Woodcuts by the Author. New Ed., small 4to, cloth extra, 6s.

Chaucer for Schools. By Mrs. H. R. HAWEIS. Demy 8vo, cloth limp, 2s. 6d.

City (The) of Dream: A Poem.

Fcap. 8vo, cloth extra, 6s. [*In the press.*

Cobban.—The Cure of Souls:

A Story. By J. MACLAREN COBBAN. Post 8vo, illustrated boards, 2s.

Collins (C. Allston).—The Bar

Sinister: A Story. By C. ALLSTON COLLINS. Post 8vo, illustrated boards, 2s.

Collins (Mortimer & Frances),

Novels by:

Sweet and Twenty. Post 8vo, illustrated boards, 2s.

Frances. Post 8vo, illust. bds., 2s.

Blacksmith and Scholar. Post 8vo, illustrated boards, 2s.; crown 8vo cloth extra, 3s. 6d.

The Village Comedy. Post 8vo, illus'. boards, 2s.; cr. 8vo, cloth extra, 3s. 6d

You Play Me False. Post 8vo, illust. boards, 2s.; cr. 8vo, cloth extra, 3s. 6d.

Collins (Mortimer), Novels by:

Sweet Anne Page. Post 8vo, illustrated boards, 2s.; crown 8vo, cloth extra, 3s. 6d.

Transmigration. Post 8vo, illustrated boards, 2s.; crown 8vo, cloth extra, 3s. 6d.

From Midnight to Midnight. Post 8vo, illustrated boards, 2s.; crown 8vo, cloth extra, 3s. 6d.

A Fight with Fortune. Post 8vo, illustrated boards, 2s.

Colman's Humorous Works:

" Broad Grins," " My Nightgown and Slippers," and other Humorous Works, Prose and Poetical, of GEORGE COLMAN. With Life by G. B BUCKSTONE, and Frontispiece by HOGARTH. Crown 8vo, cloth extra, gilt, 7s. 6d.

Collins (Wilkie), Novels by.

Each post 8vo, illustrated boards, 2s; cloth limp, 2s. 6d.; or crown 8vo, cloth extra, Illustrated, 3s. 6d.

Antonina. Illust. by A. CONCANEN.

Basil. Illustrated by Sir JOHN GILBERT and J. MAHONEY.

Hide and Seek. Illustrated by Sir JOHN GILBERT and J. MAHONEY.

The Dead Secret. Illustrated by Sir JOHN GILBERT and A. CONCANEN.

Queen of Hearts Illustrated by Sir JOHN GILBERT and A. CONCANEN.

My Miscellanies. With Illustrations by A. CONCANEN, and a Steel-plate Portrait of WILKIE COLLINS.

The Woman in White. With Illustrations by Sir JOHN GILBERT and F. A. FRASER.

The Moonstone. With Illustrations by G. DU MAURIER and F. A. FRASER.

Man and Wife. Illust. by W. SMALL.

Poor Miss Finch. Illustrated by G. DU MAURIER and EDWARD HUGHES.

Miss or Mrs.? With Illustrations by S. L. FILDES and HENRY WOODS.

The New Magdalen. Illustrated by G. DU MAURIER and C. S. RANDS.

The Frozen Deep. Illustrated by G. DU MAURIER and J. MAHONEY.

The Law and the Lady. Illustrated by S. L. FILDES and SYDNEY HALL.

The Two Destinies.

The Haunted Hotel. Illustrated by 'ARTHUR HOPKINS.

The Fallen Leaves.

Jezebel's Daughter.

The Black Robe.

Heart and Science: A Story of the Present Time. New and Cheaper Edition. Crown 8vo, cloth extra, 3s. 6d.

Convalescent Cookery: A

Family Handbook. By CATHERINE RYAN. Post 8vo, cloth limp, 2s. 6d.

Conway (Moncure D.), Works by:

Demonology and Devil-Lore. Two Vols., royal 8vo, with 65 Illusts., 28s.

A Necklace of Stories. Illustrated by W. J. HENNESSY. Square 8vo, cloth extra, 6s.

The Wandering Jew. Crown 8vo, cloth extra, 6s.

Thomas Carlyle: Letters and Re-collections. With Illustrations. Crown 8vo, cloth extra, 6s.

Cook (Dutton), Works by:

Hours with the Players. With a Steel Plate Frontispiece. New and Cheaper Edit., cr. 8vo, cloth extra, 6s.

Nights at the Play: A View of the English Stage. New and Cheaper Edition. Crown 8vo, cloth extra, 6s.

Leo: A Novel. Post 8vo, illustrated boards, 2s.

Paul Foster's Daughter. Post 8vo, illustrated boards, 2s.; crown 8vo, cloth extra, 3s. 6d.

Copyright. — A Handbook of

English and Foreign Copyright in Literary and Dramatic Works. By SIDNEY JERROLD, of the Middle Temple, Esq., Barrister-at-Law. Post 8vo, cloth limp, 2s. 6d.

Cornwall.—Popular Romances

of the West of England; or, The Drolls, Traditions, and Superstitions of Old Cornwall. Collected and Edited by ROBERT HUNT, F.R.S. New and Revised Edition, with Additions, and Two Steel-plate Illustrations by GEORGE CRUIKSHANK. Crown 8vo, cloth extra, 7s. 6d.

Creasy.—Memoirs of Eminent

Etonians: with Notices of the Early History of Eton College. By Sir EDWARD CREASY, Author of "The Fifteen Decisive Battles of the World." Crown 8vo, cloth extra, gilt, with 13 Portraits, 7s. 6d.

Cruikshank (George):

The Comic Almanack. Complete in Two SERIES: The FIRST from 1835 to 1843; the SECOND from 1844 to 1853. A Gathering of the BEST HUMOUR of THACKERAY, HOOD, MAYHEW, ALBERT SMITH, A'BECKETT, ROBERT BROUGH, &c. With 2,000 Woodcuts and Steel Engravings by CRUIKSHANK, HINE, LANDELLS, &c. Crown 8vo, cloth gilt, two very thick volumes, 7s. 6d. each.

The Life of George Cruikshank. By BLANCHARD JERROLD, Author of "The Life of Napoleon III.," &c. With 84 Illustrations. New and Cheaper Edition, enlarged, with Additional Plates, and a very carefully compiled Bibliography. Crown 8vo, cloth extra, 7s. 6d.

Robinson Crusoe. A choicely-printed Edition, with 37 Woodcuts and Two Steel Plates by GEORGE CRUIKSHANK. Crown 8vo, cloth extra, 7s. 6d. 100 Large Paper copies, carefully printed on hand-made paper, with India proofs of the Illustrations, price 36s.

Cussans.—Handbook of Heraldry; with Instructions for Tracing Pedigrees and Deciphering Ancient MSS., &c. By JOHN E. CUSSANS. Entirely New and Revised Edition, illustrated with over 400 Woodcuts and Coloured Plates. Crown 8vo, cloth extra, 7s. 6d.

Cyples.—Hearts of Gold: A Novel. By WILLIAM CYPLES. Crown 8vo, cloth extra, 3s. 6d.

Daniel. — Merrie England in the Olden Time. By GEORGE DANIEL. With Illustrations by ROBT. CRUIKSHANK. Crown 8vo, cloth extra, 3s. 6d.

Daudet.—Port Salvation; or, The Evangelist. By ALPHONSE DAUDET. Translated by C. HARRY MELTZER. With Portrait of the Author. Crown 8vo, cloth extra, 3s. 6d.

Davenant. — What shall my Son be? Hints for Parents on the Choice of a Profession or Trade for their Sons. By FRANCIS DAVENANT, M.A. Post 8vo, cloth limp, 2s. 6d.

Davies (Dr. N. E.), Works by:
One Thousand Medical Maxims. Crown 8vo, 1s.; cloth, 1s. 6d.
Nursery Hints: A Mother's Guide. Crown 8vo, 1s.; cloth, 1s. 6d.

Davies' (Sir John) Complete Poetical Works, including Psalms I. to L. in Verse, and other hitherto Unpublished MSS., for the first time Collected and Edited, with Memorial-Introduction and Notes, by the Rev. A. B. GROSART, D.D. Two Vols., crown 8vo, cloth boards, 12s.

De Maistre.—A Journey Round My Room. By XAVIER DE MAISTRE. Translated by HENRY ATTWELL. Post 8vo, cloth limp, 2s. 6d.

De Mille.—A Castle in Spain. A Novel. By JAMES DE MILLE. With a Frontispiece. Crown 8vo, cloth extra, 3s. 6d.

Derwent (Leith), Novels by:
Our Lady of Tears. Cr. 8vo, cloth extra, 3s. 6d.; post 8vo, illust. bds., 2s.
Circe's Lovers. Crown 8vo, cloth extra, 3s. 6d.

Dickens (Charles), Novels by:
Post 8vo, illustrated boards, 2s. each.
Sketches by Boz. | Nicholas Nickleby.
Pickwick Papers. | Oliver Twist.

The Speeches of Charles Dickens. (*Mayfair Library.*) Post 8vo, cloth limp, 2s. 6d.

The Speeches of Charles Dickens, 1841–1870. With a New Bibliography, revised and enlarged. Edited and Prefaced by RICHARD HERNE SHEPHERD. Crown 8vo, cloth extra, 6s.

About England with Dickens. By ALFRED RIMMER. With 57 Illustrations by C. A. VANDERHOOF, ALFRED RIMMER, and others. Sq. 8vo, cloth extra, 10s. 6d.

Dictionaries:

A Dictionary of Miracles: Imitative, Realistic, and Dogmatic. By the Rev. E. C. BREWER, LL.D. Crown 8vo, cloth extra, 7s. 6d. [*Immediately.*

A Dictionary of the Drama: Being a comprehensive Guide to the Plays, Playwrights, Players, and Playhouses of the United Kingdom and America, from the Earliest to the Present Times. By W. DAVENPORT ADAMS. A thick volume, crown 8vo, half-bound, 12s. 6d. [*In preparation.*

Familiar Allusions: A Handbook of Miscellaneous Information; including the Names of Celebrated Statues, Paintings, Palaces, Country Seats, Ruins, Churches, Ships, Streets, Clubs, Natural Curiosities, and the like. By WM. A. WHEELER and CHARLES G. WHEELER. Demy 8vo, cloth extra, 7s. 6d.

The Reader's Handbook of Allusions, References, Plots, and Stories. By the Rev. E. C. BREWER, LL.D. Third Edition, revised throughout, with a New Appendix, containing a Complete English Bibliography. Crown 8vo, 1,400 pages, cloth extra, 7s. 6d.

Short Sayings of Great Men. With Historical and Explanatory Notes. By SAMUEL A. BENT, M.A. Demy 8vo, cloth extra, 7s. 6d.

The Slang Dictionary: Etymological, Historical, and Anecdotal. Crown 8vo, cloth extra, 6s. 6d.

Words, Facts, and Phrases: A Dictionary of Curious, Quaint, and Out-of-the-Way Matters. By ELIEZER EDWARDS. Crown 8vo, half-bound, 12s. 6d.

Dobson (W. T.), Works by :

Literary Frivolities, Fancies, Follies, and Frolics. Post 8vo, cloth limp, 2s. 6d.

Poetical Ingenuities and Eccentricities. Post 8vo, cloth limp, 2s. 6d.

Doran. — **Memories of our Great Towns**; with Anecdotic Gleanings concerning their Worthies and their Oddities. By Dr. JOHN DORAN, F.S.A. With 38 Illustrations. New and Cheaper Edition, crown 8vo, cloth extra, 7s. 6d.

Drama, A Dictionary of the. Being a comprehensive Guide to the Plays, Playwrights, Players, and Playhouses of the United Kingdom and America, from the Earliest to the Present Times. By W. DAVENPORT ADAMS. (Uniform with BREWER'S "Reader's Handbook.") Crown 8vo, half-bound, 12s. 6d. [*In preparation.*

Dramatists, The Old. Crown 8vo, cloth extra, with Vignette Portraits, 6s. per Vol.

Ben Jonson's Works. With Notes Critical and Explanatory, and a Biographical Memoir by WM. GIFFORD. Edited by Colonel CUNNINGHAM. Three Vols.

Chapman's Works. Complete in Three Vols. Vol. I. contains the Plays complete, including the doubtful ones; Vol. II., the Poems and Minor Translations, with an Introductory Essay by ALGERNON CHAS. SWINBURNE; Vol. III., the Translations of the Iliad and Odyssey.

Marlowe's Works. Including his Translations. Edited, with Notes and Introduction, by Col. CUNNINGHAM. One Vol.

Massinger's Plays. From the Text of WILLIAM GIFFORD. Edited by Col. CUNNINGHAM. One Vol.

Dyer. — **The Folk-Lore of Plants.** By T. F. THISELTON DYER, M.A., &c. Crown 8vo, cloth extra, 7s. 6d. [*In preparation.*

Edwardes (Mrs. A.), Novels by :

A Point of Honour. Post 8vo, illustrated boards, 2s.

Archie Lovell. Post 8vo, illust. bds., 2s. ; crown 8vo, cloth extra, 3s. 6d.

Eggleston. — **Roxy: A Novel.** By EDWARD EGGLESTON. Post 8vo, illust. boards, 2s. ; cr. 8vo, cloth extra, 3s. 6d.

Early English Poets. Edited, with Introductions and Annotations, by Rev. A. B. GROSART, D.D. Crown 8vo, cloth boards, 6s. per Volume.

Fletcher's (Giles, B.D.) Complete Poems. One Vol.

Davies' (Sir John) Complete Poetical Works. Two Vols.

Herrick's (Robert) Complete Collected Poems. Three Vols.

Sidney's (Sir Philip) Complete Poetical Works. Three Vols.

Herbert (Lord) of Cherbury's Poems. Edited, with Introduction, by J. CHURTON COLLINS. Crown 8vo, parchment, 8s.

Emanuel. — **On Diamonds and Precious Stones:** their History, Value, and Properties ; with Simple Tests for ascertaining their Reality. By HARRY EMANUEL, F.R.G.S. With numerous Illustrations, tinted and plain. Crown 8vo, cloth extra, gilt, 6s.

Englishman's House, The : A Practical Guide to all interested in Selecting or Building a House, with full Estimates of Cost, Quantities, &c. By C. J. RICHARDSON. Third Edition. With nearly 600 Illustrations. Crown 8vo, cloth extra, 7s. 6d.

Ewald (Alex. Charles. F.S.A.), Works by :

Stories from the State Papers. With an Autotype Facsimile. Crown 8vo, cloth extra, 6s.

The Life and Times of Prince Charles Stuart, Count of Albany, commonly called the Young Pretender. From the State Papers and other Sources. New and Cheaper Edition, with a Portrait, crown 8vo, cloth extra, 7s. 6d.

Eyes, The. — **How to Use our Eyes,** and How to Preserve Them. By JOHN BROWNING, F.R.A.S., &c. With 37 Illustrations. Crown 8vo, 1s.; cloth 1s. 6d.

Fairholt. — **Tobacco :** Its History and Associations; with an Account of the Plant and its Manufacture, and its Modes of Use in all Ages and Countries. By F. W. FAIRHOLT, F.S.A. With Coloured Frontispiece and upwards of 100 Illustrations by the Author. Crown 8vo, cloth extra, 6s.

Familiar Allusions: A Handbook of Miscellaneous Information; including the Names of Celebrated Statues, Paintings, Palaces, Country Seats, Ruins, Churches, Ships, Streets, Clubs, Natural Curiosities, and the like. By WILLIAM A. WHEELER, Author of " Noted Names of Fiction ;" and CHARLES G. WHEELER. Demy 8vo, cloth extra, 7s. 6d.

Faraday (Michael), Works by :

The Chemical History of a Candle : Lectures delivered before a Juvenile Audience at the Royal Institution. Edited by WILLIAM CROOKES, F.C.S. Post 8vo, cloth extra, with numerous Illustrations, 4s. 6d.

On the Various Forces of Nature, and their Relations to each other : Lectures delivered before a Juvenile Audience at the Royal Institution. Edited by WILLIAM CROOKES, F.C.S. Post 8vo, cloth extra, with numerous Illustrations, 4s. 6d.

Fin-Bec. — The Cupboard Papers: Observations on the Art of Living and Dining. By FIN-BEC. Post 8vo, cloth limp, 2s. 6d.

Fitzgerald (Percy), Works by :

The Recreations of a Literary Man ; or, Does Writing Pay? With Recollections of some Literary Men, and a View of a Literary Man's Working Life. Cr. 8vo, cloth extra, 6s.

The World Behind the Scenes. Crown 8vo, cloth extra, 3s. 6d.

Little Essays: Passages from the Letters of CHARLES LAMB. Post 8vo, cloth limp, 2s. 6d.

Post 8vo, illustrated boards, 2s. each.

Bella Donna. | Never Forgotten.

The Second Mrs. Tillotson.

Polly.

Seventy-five Brooke Street.

Fletcher's (Giles, B.D.) Complete Poems: Christ's Victorie in Heaven, Christ's Victorie on Earth, Christ's Triumph over Death, and Minor Poems. With Memorial-Introduction and Notes by the Rev. A. B. GROSART, D.D. Cr. 8vo, cloth bds., 6s.

Fonblanque.—Filthy Lucre : A Novel. By ALBANY DE FONBLANQUE. Post 8vo, illustrated boards, 2s.

French Literature, History of. By HENRY VAN LAUN. Complete in 3 Vols., demy 8vo, cl. bds., 7s. 6d. each.

Francillon (R. E.), Novels by :

Crown 8vo, cloth extra, 3s. 6d. each ; post 8vo, illust. boards, 2s. each.

Olympia. | Queen Cophetua.

One by One.

Esther's Glove. Fcap. 8vo, picture cover, 1s.

A Real Queen. Three Vols., cr. 8vo.

Frere.—Pandurang Hari ; or, Memoirs of a Hindoo. With a Preface by Sir H. BARTLE FRERE, G.C.S.I., &c. Crown 8vo, cloth extra, 3s. 6d. ; post 8vo, illustrated boards, 2s.

Friswell.—One of Two: A Novel. By HAIN FRISWELL. Post 8vo, illustrated boards, 2s.

Frost (Thomas), Works by :

Crown 8vo, cloth extra, 3s. 6d. each.

Circus Life and Circus Celebrities.

The Lives of the Conjurers.

The Old Showmen and the Old London Fairs.

Fry.—Royal Guide to the London Charities, 1884-5. By HERBERT FRY. Showing, in alphabetical order, their Name, Date of Foundation, Address, Objects, Annual Income, Chief Officials, &c. Published Annually. Crown 8vo, cloth, 1s. 6d. [*Immediately.*

Gardening Books:

A Year's Work in Garden and Greenhouse: Practical Advice to Amateur Gardeners as to the Management of the Flower, Fruit, and Frame Garden. By GEORGE GLENNY. Post 8vo, cloth limp, 2s. 6d.

Our Kitchen Garden: The Plants we Grow, and How we Cook Them. By TOM JERROLD, Author of " The Garden that Paid the Rent," &c. Post 8vo, cloth limp, 2s. 6d.

Household Horticulture: A Gossip about Flowers. By TOM and JANE JERROLD. Illustrated. Post 8vo, cloth limp, 2s. 6d.

The Garden that Paid the Rent. By TOM JERROLD. Fcap. 8vo, illustrated cover, 1s. ; cloth limp, 1s. 6d.

Garrett.—The Capel Girls: A Novel. By EDWARD GARRETT. Post 8vo, illust. bds., 2s. ; cr. 8vo, cl. ex., 3s. 6d.

German Popular Stories. Collected by the Brothers GRIMM, and Translated by EDGAR TAYLOR. Edited, with an Introduction, by JOHN RUSKIN. With 22 Illustrations on Steel by GEORGE CRUIKSHANK. Square 8vo, cloth extra, 6s. 6d. gilt edges, 7s. 6d.

Gentleman's Magazine (The)

for 1884. One Shilling Monthly. A New Serial Story, entitled "Philistia," by CECIL POWER, is now appearing. "Science Notes," by W. MATTIEU WILLIAMS, F.R.A.S., and "Table Talk," by SYLVANUS URBAN, are also continued monthly.

*** *Now ready, the Volume for* JULY *to* DECEMBER, 1883, *cloth extra, price* 8s. 6d.; *Cases for binding,* 2s. *each.*

Gibbon (Charles), Novels by:

Crown 8vo, cloth extra, 3s. 6d. each; post 8vo, illustrated boards, 2s. each.

Robin Gray.
For Lack of Gold.
What will the World Say?
In Honour Bound.
In Love and War.
For the King.
Queen of the Meadow.
In Pastures Green.
The Braes of Yarrow.
The Flower of the Forest.
A Heart's Problem.

Post 8vo, illustrated boards, 2s.
The Dead Heart.

Crown 8vo, cloth extra, 3s. 6d. each.
The Golden Shaft.
Of High Degree.

Fancy-Free. Three Vols., crown 8vo.

Gilbert (William), Novels by:

Post 8vo, illustrated boards, 2s. each.

Dr. Austin's Guests.
The Wizard of the Mountain.
James Duke, Costermonger.

Gilbert (W. S.), Original Plays

by: In Two Series, each complete in itself, price 2s. 6d. each.

The FIRST SERIES contains — The Wicked World—Pygmalion and Galatea — Charity — The Princess — The Palace of Truth—Trial by Jury.

The SECOND SERIES contains—Broken Hearts—Engaged—Sweethearts—Gretchen—Dan'l Druce—Tom Cobb—H.M.S. Pinafore—The Sorcerer—The Pirates of Penzance.

Glenny.—A Year's Work in

Garden and Greenhouse: Practical Advice to Amateur Gardeners as to the Management of the Flower, Fruit, and Frame Garden. By GEORGE GLENNY. Post 8vo, cloth limp, 2s. 6d.

Godwin.—Lives of the Necromancers.

By WILLIAM GODWIN. Post 8vo, cloth limp, 2s.

Golden Library, The:

Square 16mo (Tauchnitz size), cloth limp, 2s. per volume.

Bayard Taylor's Diversions of the Echo Club.

Bennett's (Dr. W. C.) Ballad History of England.

Bennett's (Dr. W. C.) Songs for Sailors.

Byron's Don Juan.

Godwin's (William) Lives of the Necromancers.

Holmes's Autocrat of the Breakfast Table. With an Introduction by G. A. SALA.

Holmes's Professor at the Breakfast Table.

Hood's Whims and Oddities. Complete. All the original Illustrations.

Irving's (Washington) Tales of a Traveller.

Irving's (Washington) Tales of the Alhambra.

Jesse's (Edward) Scenes and Occupations of a Country Life.

Lamb's Essays of Elia. Both Series Complete in One Vol.

Leigh Hunt's Essays: A Tale for a Chimney Corner, and other Pieces. With Portrait, and Introduction by EDMUND OLLIER.

Mallory's (Sir Thomas) Mort d'Arthur: The Stories of King Arthur and of the Knights of the Round Table. Edited by B. MONTGOMERIE RANKING.

Pascal's Provincial Letters. A New Translation, with Historical Introduction and Notes, by T. M'CRIE, D.D.

Pope's Poetical Works. Complete.

Rochefoucauld's Maxims and Moral Reflections. With Notes, and Introductory Essay by SAINTE-BEUVE.

St. Pierre's Paul and Virginia, and The Indian Cottage. Edited, with Life, by the Rev. E. CLARKE.

Shelley's Early Poems, and Queen Mab. With Essay by LEIGH HUNT.

Shelley's Later Poems: Laon and Cythna, &c.

Shelley's Posthumous Poems, the Shelley Papers, &c.

Shelley's Prose Works, including A Refutation of Deism, Zastrozzi, St. Irvyne, &c.

White's Natural History of Selborne. Edited, with Additions, by THOMAS BROWN, F.L.S.

Golden Treasury of Thought,
The: An ENCYCLOPÆDIA OF QUOTA-
TIONS from Writers of all Times and
Countries. Selected and Edited by
THEODORE TAYLOR. Crown 8vo, cloth
gilt and gilt edges, 7s. 6d.

Gordon Cumming(C. F.),Works
by:

In the Hebrides. With Autotype Facsimile and numerous full-page Illustrations. Demy 8vo, cloth extra,
8s. 6d.

In the Himalayas. With numerous
Illustrations. Demy 8vo, cloth extra,
8s. 6d. [*Shortly.*

Graham. — The Professor's
Wife: A Story. By LEONARD GRAHAM.
Fcap. 8vo, picture cover, 1s.; cloth
extra, 2s. 6d.

Greeks and Romans, The Life
of the, Described from Antique Monu-
ments. By ERNST GUHL and W.
KONER. Translated from the Third
German Edition, and Edited by Dr.
F. HUEFFER. With 545 Illustrations.
New and Cheaper Edition, demy 8vo,
cloth extra, 7s. 6d.

Greenwood (James),Works by:

The Wilds of London. Crown 8vo,
cloth extra, 3s. 6d.

Low-Life Deeps: An Account of the
Strange Fish to be Found There.
Crown 8vo, cloth extra, 3s. 6d.

Dick Temple: A Novel. Post 8vo,
illustrated boards, 2s.

Guyot.—The Earth and Man;
or, Physical Geography in its relation
to the History of Mankind. By
ARNOLD GUYOT. With Additions by
Professors AGASSIZ, PIERCE, and GRAY;
12 Maps and Engravings on Steel,
some Coloured, and copious Index.
Crown 8vo, cloth extra, gilt, 4s. 6d.

Hair (The): Its Treatment in
Health, Weakness, and Disease.
Translated from the German of Dr. J.
PINCUS. Crown 8vo, 1s.; cloth, 1s. 6d.

Hake (Dr. Thomas Gordon),
Poems by:

Maiden Ecstasy. Small 4to, cloth
extra, 8s.

New Symbols. Crown 8vo, cloth
extra, 6s.

Legends of the Morrow. Crown 8vo,
cloth extra, 6s.

The Serpent Play. Crown 8vo, cloth
extra, 6s.

Hall.—Sketches of Irish Cha-
racter. By Mrs. S. C. HALL. With
numerous Illustrations on Steel and
Wood by MACLISE, GILBERT, HARVEY,
and G. CRUIKSHANK. Medium 8vo,
cloth extra, gilt, 7s. 6d.

Halliday.—Every-day Papers.
By ANDREW HALLIDAY. Post 8vo,
illustrated boards, 2s.

Handwriting, The Philosophy
of. With over 100 Facsimiles and Explanatory Text. By DON FELIX DE
SALAMANCA. Post 8vo, cloth limp,
2s. 6d.

Hanky-Panky: A Collection of
Very EasyTricks,Very Difficult Tricks,
White Magic, Sleight of Hand, &c.
Edited by W. H. CREMER. With 200
Illustrations. Crown 8vo, cloth extra,
4s. 6d.

Hardy (Lady Duffus). — Paul
Wynter's Sacrifice: A Story. By
Lady DUFFUS HARDY. Post 8vo, illust.
boards, 2s.

Hardy (Thomas).—Under the
Greenwood Tree. By THOMAS HARDY,
Author of "Far from the Madding
Crowd." Crown 8vo, cloth extra,
3s. 6d.; post 8vo, illustrated boards,
2s.

Haweis (Mrs. H. R.), Works by:

The Art of Dress. With numerous
Illustrations. Small 8vo, illustrated
cover, 1s.; cloth limp, 1s. 6d.

The Art of Beauty. New and Cheaper
Edition. Crown 8vo, cloth extra,
with Coloured Frontispiece and Illustrations, 6s.

The Art of Decoration. Square 8vo,
handsomely bound and profusely
Illustrated, 10s. 6d.

Chaucer for Children: A Golden
Key. With Eight Coloured Pictures
and numerous Woodcuts. New
Edition, small 4to, cloth extra, 6s.

Chaucer for Schools. Demy 8vo,
cloth limp, 2s. 6d.

Haweis(Rev. H. R.).—American
Humorists. Including WASHINGTON
IRVING, OLIVER WENDELL HOLMES,
JAMES RUSSELL LOWELL, ARTEMUS
WARD,MARK TWAIN, and BRET HARTE.
By the Rev. H. R. HAWEIS, M.A.
Crown 8vo, cloth extra, 6s.

Hawthorne (Julian), Novels by.
Crown 8vo, cloth extra, 3s. 6d. each ;
post 8vo, illustrated boards, 2s. each.

Garth.

Ellice Quentin.

Sebastian Strome.

Prince Saroni's Wife.

Dust.

Mrs. Gainsborough's Diamonds.
Fcap. 8vo, illustrated cover, 1s. ;
cloth extra, 2s. 6d.

Fortune's Fool. Crown 8vo, cloth
extra, 3s. 6d.

Beatrix Randolph. With Illustrations
by A. FREDERICKS. Crown 8vo, cloth
extra, 3s. 6d. [Preparing.

Heath (F. G.). — My Garden
Wild, and What I Grew There. By
FRANCIS GEORGE HEATH, Author of
"The Fern World," &c. Crown 8vo,
cloth extra, 5s. ; cloth gilt, and gilt
edges, 6s.

Helps (Sir Arthur), Works by :
Animals and their Masters. Post
8vo, cloth limp, 2s. 6d.

Social Pressure. Post 8vo, cloth limp,
2s. 6d.

Ivan de Biron: A Novel. Crown 8vo,
cloth extra, 3s. 6d.; post 8vo, illustrated boards, 2s.

Heptalogia (The); or, The
Seven against Sense. A Cap with
Seven Bells. Cr. 8vo, cloth extra, 6s.

Herbert.—The Poems of Lord
Herbert of Cherbury. Edited, with
an Introduction, by J. CHURTON
COLLINS. Crown 8vo, bound in parchment, 8s.

Herrick's (Robert) Hesperides,
Noble Numbers, and Complete Collected Poems. With Memorial-Introduction and Notes by the Rev. A. B.
GROSART, D.D., Steel Portrait, Index
of First Lines, and Glossarial Index,
&c. Three Vols., crown 8vo, cloth
boards, 18s.

Hesse-Wartegg (Chevalier
Ernst von), Works by :
Tunis: The Land and the People.
With 22 Illustrations. Crown 8vo,
cloth extra, 3s. 6d.

The New South-West: Travelling
Sketches from Kansas, New Mexico,
Arizona, and Northern Mexico.
With 100 fine Illustrations and Three
Maps. Demy 8vo, cloth extra,
14s. [In preparation.

Hindley (Charles), Works by :
Crown 8vo, cloth extra, 3s. 6d. each.
Tavern Anecdotes and Sayings: In-
cluding the Origin of Signs, and
Reminiscences connected with
Taverns, Coffee Houses, Clubs, &c.
With Illustrations.

The Life and Adventures of a Cheap
Jack. By One of the Fraternity.
Edited by CHARLES HINDLEY.

Holmes (O. Wendell), Works by :
The Autocrat of the Breakfast-
Table. Illustrated by J. GORDON
THOMSON. Post 8vo, cloth limp,
2s. 6d.; another Edition in smaller
type, with an Introduction by G. A.
SALA. Post 8vo, cloth limp, 2s.

The Professor at the Breakfast-
Table; with the Story of Iris. Post
8vo, cloth limp, 2s.

Holmes. — The Science of
Voice Production and Voice Preser-
vation: A Popular Manual for the
Use of Speakers and Singers. By
GORDON HOLMES, M.D. Crown 8vo,
cloth limp, with Illustrations, 2s. 6d.

Hood (Thomas):
Hood's Choice Works, in Prose and
Verse. Including the Cream of the
Comic Annuals. With Life of the
Author, Portrait, and 200 Illustra-
tions. Crown 8vo, cloth extra, 7s. 6d.

Hood's Whims and Oddities. Complete. With all the original Illustrations. Post 8vo, cloth limp, 2s.

Hood (Tom), Works by :
From Nowhere to the North Pole:
A Noah's Arkæological Narrative.
With 25 Illustrations by W. BRUN-
TON and E. C. BARNES. Square
crown 8vo, cloth extra, gilt edges, 6s.

A Golden Heart: A Novel. Post 8vo,
illustrated boards, 2s.

Hook's (Theodore) Choice Humorous Works, including his Ludi-
crous Adventures, Bons Mots, Puns and
Hoaxes. With a New Life of the
Author, Portraits, Facsimiles, and
Illustrations. Crown 8vo, cloth extra,
gilt, 7s. 6d.

Hooper.—The House of Raby :
A Novel. By Mrs. GEORGE HOOPER.
Post 8vo, illustrated boards, 2s.

Horne.—Orion : An Epic Poem,
in Three Books. By RICHARD HEN-
GIST HORNE. With Photographic
Portrait from a Medallion by SUM-
MERS. Tenth Edition, crown 8vo,
cloth extra, 7s.

Howell.—Conflicts of Capital and Labour, Historically and Economically considered: Being a History and Review of the Trade Unions of Great Britain, showing their Origin, Progress, Constitution, and Objects, in their Political, Social, Economical, and Industrial Aspects. By GEORGE HOWELL. Cr. 8vo, cloth extra, 7s. 6d.

Hugo. — The Hunchback of Notre Dame. By VICTOR HUGO. Post 8vo, illustrated boards, 2s.

Hunt.—Essays by Leigh Hunt. A Tale for a Chimney Corner, and other Pieces. With Portrait and Introduction by EDMUND OLLIER. Post 8vo, cloth limp, 2s.

Hunt (Mrs. Alfred), Novels by: Crown 8vo, cloth extra, 3s. 6d. each; post 8vo, illustrated boards, 2s. each.
    Thornicroft's Model.
    The Leaden Casket.
    Self-Condemned.

Ingelow.—Fated to be Free: A Novel. By JEAN INGELOW. Crown 8vo, cloth extra, 3s. 6d.; post 8vo, illustrated boards, 2s.

Irish Wit and Humour, Songs of. Collected and Edited by A. PERCEVAL GRAVES. Post 8vo, cloth limp, 2s. 6d.

Irving (Henry).—The Paradox of Acting. Translated, with Annotations, from Diderot's "Le Paradoxe sur le Comédien," by WALTER HERRIES POLLOCK. With a Preface by HENRY IRVING. Crown 8vo, in parchment, 4s. 6d.

Irving (Washington),Works by: Post 8vo, cloth limp, 2s. each.
    Tales of a Traveller.
    Tales of the Alhambra.

James.—Confidence: A Novel. By HENRY JAMES, Jun. Crown 8vo, cloth extra, 3s. 6d.; post 8vo, illustrated boards, 2s.

Janvier.—Practical Keramics for Students. By CATHERINE A. JANVIER. Crown 8vo, cloth extra, 6s.

Jay (Harriett), Novels by. Each crown 8vo, cloth extra, 3s. 6d.; or post 8vo, illustrated boards, 2s.
    The Dark Colleen.
    The Queen of Connaught.

Jefferies (Richard), Works by:
Nature near London. Crown 8vo, cloth extra, 6s.
The Life of the Fields. Crown 8vo, cloth extra, 6s. [*In the press.*

Jennings (H. J.).—Curiosities of Criticism. By HENRY J. JENNINGS. Post 8vo, cloth limp, 2s. 6d.

Jennings (Hargrave). — The Rosicrucians: Their Rites and Mysteries. With Chapters on the Ancient Fire and Serpent Worshippers. By HARGRAVE JENNINGS. With Five full-page Plates and upwards of 300 Illustrations. A New Edition, crown 8vo, cloth extra, 7s. 6d.

Jerrold (Tom), Works by:
The Garden that Paid the Rent By TOM JERROLD. Fcap. 8vo, illustrated cover, 1s.; cloth limp, 1s. 6d.
Household Horticulture: A Gossip about Flowers. By TOM and JANE JERROLD. Illustrated. Post 8vo, cloth limp, 2s. 6d.
Our Kitchen Garden: The Plants we Grow, and How we Cook Them. By TOM JERROLD. Post 8vo, cloth limp, 2s. 6d.

Jesse.—Scenes and Occupations of a Country Life. By EDWARD JESSE. Post 8vo, cloth limp, 2s.

Jones (Wm., F.S.A.), Works by:
Finger-Ring Lore: Historical, Legendary, and Anecdotal. With over 200 Illustrations. Crown 8vo, cloth extra, 7s. 6d.
Credulities, Past and Present; including the Sea and Seamen, Miners, Talismans,Word and Letter Divination, Exorcising and Blessing of Animals, Birds, Eggs, Luck, &c. With an Etched Frontispiece. Crown 8vo, cloth extra, 7s. 6d.
Crowns and Coronations: A History of Regalia in all Times and Countries. With One Hundred Illustrations. Cr. 8vo, cloth extra, 7s. 6d.

Jonson's (Ben) Works. With Notes Critical and Explanatory, and a Biographical Memoir by WILLIAM GIFFORD. Edited by Colonel CUNNINGHAM. Three Vols., crown 8vo, cloth extra, 18s.; or separately, 6s. each.

Josephus,The Complete Works of. Translated by WHISTON. Containing both "The Antiquities of the Jews" and "The Wars of the Jews." Two Vols., 8vo, with 52 Illustrations and Maps, cloth extra, gilt, 14s.

Kavanagh.—The Pearl Fountain, and other Fairy Stories. By BRIDGET and JULIA KAVANAGH. With Thirty Illustrations by J. MOYR SMITH. Small 8vo, cloth gilt, 6s.

Kempt.—Pencil and Palette: Chapters on Art and Artists. By ROBERT KEMPT. Post 8vo, cloth limp, 2s. 6d.

Kingsley (Henry), Novels by: Each crown 8vo, cloth extra, 3s. 6d.; or post 8vo, illustrated boards, 2s.

Oakshott Castle. | Number Seventeen

Lamb (Charles):

Mary and Charles Lamb: Their Poems, Letters, and Remains. With Reminiscences and Notes by W. CAREW HAZLITT. With HANCOCK'S Portrait of the Essayist, Facsimiles of the Title-pages of the rare First Editions of Lamb's and Coleridge's Works, and numerous Illustrations. Crown 8vo, cloth extra, 10s. 6d.

Lamb's Complete Works, in Prose and Verse, reprinted from the Original Editions, with many Pieces hitherto unpublished. Edited, with Notes and Introduction, by R. H. SHEPHERD. With Two Portraits and Facsimile of Page of the "Essay on Roast Pig." Cr. 8vo, cloth extra, 7s. 6d.

The Essays of Elia. Complete Edition. Post 8vo, cloth extra, 2s.

Poetry for Children, and Prince Dorus. By CHARLES LAMB. Carefully Reprinted from unique copies. Small 8vo, cloth extra, 5s.

Little Essays: Sketches and Characters. By CHARLES LAMB. Selected from his Letters by PERCY FITZGERALD. Post 8vo, cloth limp, 2s. 6d.

Lane's Arabian Nights, &c.:

The Thousand and One Nights: commonly called, in England, "THE ARABIAN NIGHTS' ENTERTAINMENTS." A New Translation from the Arabic, with copious Notes, by EDWARD WILLIAM LANE. Illustrated by many hundred Engravings on Wood, from Original Designs by WM. HARVEY. A New Edition, from a Copy annotated by the Translator, edited by his Nephew, EDWARD STANLEY POOLE. With a Preface by STANLEY LANE-POOLE. Three Vols., demy 8vo, cloth extra, 7s. 6d. each.

Arabian Society in the Middle Ages : Studies from "The Thousand and One Nights." By EDWARD WILLIAM LANE, Author of "The Modern Egyptians," &c. Edited by STANLEY LANE-POOLE. Cr. 8vo, cloth extra, 6s.

Lares and Penates; or, The Background of Life. By FLORENCE CADDY. Crown 8vo, cloth extra, 6s.

Larwood (Jacob), Works by:

The Story of the London Parks. With Illustrations. Crown 8vo, cloth extra, 3s. 6d.

Clerical Anecdotes. Post 8vo, cloth limp, 2s. 6d.

Forensic Anecdotes Post 8vo, cloth limp, 2s. 6d.

Theatrical Anecdotes. Post 8vo, cloth limp, 2s. 6d.

Leigh (Henry S.), Works by:

Carols of Cockayne. With numerous Illustrations. Post 8vo, cloth limp, 2s. 6d.

Jeux d'Esprit. Collected and Edited by HENRY S. LEIGH. Post 8vo, cloth limp, 2s. 6d.

Life in London; or, The History of Jerry Hawthorn and Corinthian Tom. With the whole of CRUIKSHANK'S Illustrations, in Colours, after the Originals. Crown 8vo, cloth extra, 7s. 6d.

Linton (E. Lynn), Works by Post 8vo, cloth limp, 2s. 6d. each.

Witch Stories.

The True Story of Joshua Davidson.

Ourselves Essays on Women.

Crown 8vo, cloth extra, 3s. 6d. each ; post 8vo, illustrated boards, 2s. each.

Patricia Kemball.

The Atonement of Leam Dundas.

The World Well Lost.

Under which Lord ?

With a Silken Thread.

The Rebel of the Family.

"My Love!"

Ione. Three Vols., crown 8vo.

Locks and Keys.—On the De-velopment and Distribution of Primitive Locks and Keys. By Lieut.-Gen. PITT-RIVERS, F.R.S. With numerous Illustrations. Demy 4to, half Roxburghe, 16s.

Longfellow:

Longfellow's Complete Prose Works. Including "Outre Mer," "Hyperion," "Kavanagh," "The Poets and Poetry of Europe," and "Driftwood." With Portrait and Illustrations by VALENTINE BROMLEY. Crown 8vo, cloth extra, 7s. 6d.

Longfellow's Poetical Works. Carefully Reprinted from the Original Editions. With numerous fine Illustrations on Steel and Wood. Crown 8vo, cloth extra, 7s. 6d.

Lucy.—Gideon Fleyce: A Novel. By HENRY W. LUCY. Crown 8vo, cloth extra, 3s. 6d.; post 8vo, illustrated boards, 2s.

Lusiad (The) of Camoens. Translated into English Spenserian Verse by ROBERT FFRENCH DUFF. Demy 8vo, with Fourteen full-page Plates, cloth boards, 18s.

McCarthy (Justin, M.P.),Works by:

A History of Our Own Times, from the Accession of Queen Victoria to the General Election of 1880. Four Vols. demy 8vo, cloth extra, 12s. each.—Also a POPULAR EDITION, in Four Vols. crown 8vo, cloth extra, 6s. each.

A Short History of Our Own Times. One Volume, crown 8vo, cloth extra, 6s.

History of the Four Georges. Four Vols. demy 8vo, cloth extra, 12s. each. [Vol. I. *in the press.*

Crown 8vo, cloth extra, 3s. 6d. each; post 8vo, illustrated boards, 2s. each.

Dear Lady Disdain.
The Waterdale Neighbours.
My Enemy's Daughter.
A Fair Saxon.
Linley Rochford
Miss Misanthrope.
Donna Quixote.
The Comet of a Season.

Maid of Athens. With 12 Illustrations by F. BARNARD. Three Vols., crown 8vo.

McCarthy (Justin H.), Works by:
Serapion, and other Poems. Crown 8vo, cloth extra, 6s.
An Outline of the History of Ireland, from the Earliest Times to the Present Day. Cr. 8vo, 1s.; cloth, 1s. 6d.

MacDonald (George, LL.D.), Works by:
The Princess and Curdie. With 11 Illustrations by JAMES ALLEN. Small crown 8vo, cloth extra, 5s.

Gutta-Percha Willie, the Working Genius. With 9 Illustrations by ARTHUR HUGHES. Square 8vo, cloth extra, 3s. 6d.

Paul Faber, Surgeon. With a Frontispiece by J. E. MILLAIS. Crown 8vo, cloth extra, 3s. 6d.; post 8vo, illustrated boards, 2s.

Thomas Wingfold, Curate. With a Frontispiece by C. J. STANILAND. Crown 8vo, cloth extra, 3s. 6d.; post 8vo, illustrated boards, 2s.

Macdonell.—Quaker Cousins: A Novel. By AGNES MACDONELL. Crown 8vo, cloth extra, 3s. 6d.; post 8vo, illustrated boards, 2s.

Macgregor. — Pastimes and Players. Notes on Popular Games. By ROBERT MACGREGOR. Post 8vo, cloth limp, 2s. 6d.

Maclise Portrait-Gallery (The) of Illustrious Literary Characters; with Memoirs—Biographical, Critical, Bibliographical, and Anecdotal—illustrative of the Literature of the former half of the Present Century. By WILLIAM BATES, B.A. With 85 Portraits printed on an India Tint. Crown 8vo, cloth extra, 7s. 6d.

Macquoid (Mrs.), Works by:
In the Ardennes. With 50 fine Illustrations by THOMAS R. MACQUOID. Square 8vo, cloth extra, 10s. 6d.

Pictures and Legends from Normandy and Brittany. With numerous Illustrations by THOMAS R. MACQUOID. Square 8vo, cloth gilt, 10s. 6d.

Through Normandy. With 90 Illustrations by T. R. MACQUOID. Square 8vo, cloth extra, 7s. 6d.

Through Brittany. With numerous Illustrations by T. R. MACQUOID. Square 8vo, cloth extra, 7s. 6d.

About Yorkshire With 67 Illustrations by T. R. MACQUOID, Engraved by SWAIN. Square 8vo, cloth extra, 10s. 6d.

The Evil Eye, and other Stories. Crown 8vo, cloth extra, 3s. 6d.; post 8vo, illustrated boards, 2s.

Lost Rose, and other Stories. Crown 8vo, cloth extra, 3s. 6d.; post 8vo, illustrated boards, 2s.

Mackay.—Interludes and Un-
dertones: or, Music at Twilight. By
CHARLES MACKAY, LL.D. Crown 8vo,
cloth extra, 6s.

Magician's Own Book (The):
Performances with Cups and Balls,
Eggs, Hats, Handkerchiefs, &c. All
from actual Experience. Edited by
W. H. CREMER. With 200 Illustrations.
Crown 8vo, cloth extra, 4s. 6d.

Magic No Mystery: Tricks with
Cards, Dice, Balls, &c., with fully
descriptive Directions; the Art of
Secret Writing; Training of Perform-
ing Animals, &c. With Coloured
Frontispiece and many Illustrations.
Crown 8vo, cloth extra, 4s. 6d.

Magna Charta. An exact Facsimile of the Original in the British
Museum, printed on fine plate paper,
3 feet by 2 feet, with Arms and Seals
emblazoned in Gold and Colours.
Price 5s.

Mallock (W. H.), Works by:
The New Republic; or, Culture, Faith
and Philosophy in an English Country
House. Post 8vo, cloth limp, 2s. 6d.;
Cheap Edition, illustrated boards, 2s.
The New Paul and Virginia; or, Posi-
tivism on an Island. Post 8vo, cloth
limp, 2s. 6d.
Poems. Small 4to, bound in parch-
ment, 8s.
Is Life worth Living? Crown 8vo,
cloth extra, 6s.

Mallory's (Sir Thomas) Mort
d'Arthur: The Stories of King Arthur
and of the Knights of the Round Table.
Edited by B. MONTGOMERIE RANKING.
Post 8vo, cloth limp, 2s.

Marlowe's Works. Including
his Translations. Edited, with Notes
and Introduction, by Col. CUNNING-
HAM. Crown 8vo, cloth extra, 6s.

Marryat (Florence), Novels by:
Crown 8vo, cloth extra, 3s. 6d. each; or,
post 8vo, illustrated boards, 2s.
Open! Sesame!
Written in Fire.

Post 8vo, illustrated boards, 2s. each.
A Harvest of Wild Oats.
A Little Stepson.
Fighting the Air.

Masterman.—Half a Dozen
Daughters: A Novel. By J. MASTER-
MAN. Post 8vo, illustrated boards, 2s.

Mark Twain, Works by:
The Choice Works of Mark Twain.
Revised and Corrected throughout by
the Author. With Life, Portrait, and
numerous Illustrations. Crown 8vo,
cloth extra, 7s. 6d.
The Adventures of Tom Sawyer.
With 100 Illustrations. Small 8vo,
cloth extra, 7s. 6d. CHEAP EDITION,
illustrated boards, 2s.
An Idle Excursion,and other Sketches.
Post 8vo, illustrated boards, 2s.
The Prince and the Pauper. With
nearly 200 Illustrations. Crown 8vo,
cloth extra, 7s. 6d.
The Innocents Abroad; or, The New
Pilgrim's Progress: Being some Ac-
count of the Steamship "Quaker
City's" Pleasure Excursion to
Europe and the Holy Land. With
234 Illustrations. Crown 8vo, cloth
extra, 7s. 6d. CHEAP EDITION (under
the title of "MARK TWAIN'S PLEASURE
TRIP"), post 8vo, illust. boards, 2s.
A Tramp Abroad. With 314 Illustra-
tions. Crown 8vo, cloth extra, 7s. 6d.
Without Illustrations, post 8vo, illustrated boards, 2s.
The Stolen White Elephant, &c.
Crown 8vo, cloth extra, 6s.; post 8vo,
illustrated boards, 2s.
Life on the Mississippi. With about
300 Original Illustrations. Crown
8vo, cloth extra, 7s. 6d.
The Adventures of Huckleberry
Finn. With numerous Illustrations
by the Author. Crown 8vo, cloth
extra, 7s. 6d.   [Preparing.

Massinger's Plays. From the
Text of WILLIAM GIFFORD. Edited
by Col. CUNNINGHAM. Crown 8vo,
cloth extra, 6s.

Mayhew.—London Characters
and the Humorous Side of London
Life. By HENRY MAYHEW. With
numerous Illustrations. Crown 8vo,
cloth extra, 3s. 6d.

Mayfair Library, The:
Post 8vo, cloth limp, 2s. 6d. per Volume.
A Journey Round My Room. By
XAVIER DE MAISTRE. Translated
by HENRY ATTWELL.
Latter-Day Lyrics. Edited by W.
DAVENPORT ADAMS.
Quips and Quiddities. Selected by
W. DAVENPORT ADAMS.
The Agony Column of "The Times,"
from 1800 to 1870. Edited, with an
Introduction, by ALICE CLAY.
Balzac's "Comedie Humaine" and
its Author. With Translations by
H. H. WALKER.

MAYFAIR LIBRARY, *continued—*

Melancholy Anatomised: A Popular Abridgment of "Burton's Anatomy of Melancholy."

Gastronomy as a Fine Art. By BRILLAT-SAVARIN.

The Speeches of Charles Dickens.

Literary Frivolities, Fancies, Follies, and Frolics. By W. T. DOBSON.

Poetical Ingenuities and Eccentricities. Selected and Edited by W. T. DOBSON.

The Cupboard Papers. By FIN-BEC.

Original Plays by W. S. GILBERT. FIRST SERIES. Containing: The Wicked World — Pygmalion and Galatea— Charity — The Princess— The Palace of Truth—Trial by Jury.

Original Plays by W. S. GILBERT. SECOND SERIES. Containing: Broken Hearts — Engaged — Sweethearts— Gretchen—Dan'l Druce—Tom Cobb —H.M.S. Pinafore — The Sorcerer —The Pirates of Penzance.

Songs of Irish Wit and Humour. Collected and Edited by A. PERCEVAL GRAVES.

Animals and their Masters. By Sir ARTHUR HELPS.

Social Pressure. By Sir A. HELPS.

Curiosities of Criticism. By HENRY J. JENNINGS.

The Autocrat of the Breakfast-Table. By OLIVER WENDELL HOLMES. Illustrated by J. GORDON THOMSON.

Pencil and Palette. By ROBERT KEMPT.

Little Essays: Sketches and Characters. By CHARLES LAMB. Selected from his Letters by PERCY FITZGERALD.

Clerical Anecdotes. By JACOB LARWOOD.

Forensic Anecdotes; or, Humour and Curiosities of the Law and Men of Law. By JACOB LARWOOD.

Theatrical Anecdotes. By JACOB LARWOOD.

Carols of Cockayne. By HENRY S. LEIGH.

Jeux d'Esprit. Edited by HENRY S. LEIGH.

True History of Joshua Davidson. By E. LYNN LINTON.

Witch Stories. By E. LYNN LINTON.

Ourselves: Essays on Women. By E. LYNN LINTON.

Pastimes and Players. By ROBERT MACGREGOR.

The New Paul and Virginia. By W. H. MALLOCK.

MAYFAIR LIBRARY, *continued—*

The New Republic. By W. H. MALLOCK.

Puck on Pegasus. By H. CHOLMONDELEY-PENNELL.

Pegasus Re-Saddled. By H. CHOLMONDELEY-PENNELL. Illustrated by GEORGE DU MAURIER.

Muses of Mayfair. Edited by H. CHOLMONDELEY-PENNELL.

Thoreau: His Life and Aims. By H. A. PAGE.

Puniana. By the Hon. HUGH ROWLEY.

More Puniana. By the Hon. HUGH ROWLEY.

The Philosophy of Handwriting. By DON FELIX DE SALAMANCA.

By Stream and Sea. By WILLIAM SENIOR.

Old Stories Re-told. By WALTER THORNBURY.

Leaves from a Naturalist's Note-Book. By Dr. ANDREW WILSON.

Medicine, Family.—One Thousand Medical Maxims and Surgical Hints, for Infancy, Adult Life, Middle Age, and Old Age. By N. E. DAVIES, Licentiate of the Royal College of Physicians of London. Crown 8vo, 1s.; cloth, 1s. 6d.

Merry Circle (The): A Book of New Intellectual Games and Amusements. By CLARA BELLEW. With numerous Illustrations. Crown 8vo, cloth extra, 4s. 6d.

Middlemass (Jean), Novels by:
Touch and Go. Crown 8vo, cloth extra, 3s. 6d.; post 8vo, illust. bds., 2s.

Mr. Dorillion. Post 8vo, illust. bds., 2s.

Miller. — Physiology for the Young; or, The House of Life: Human Physiology, with its application to the Preservation of Health. For use in Classes and Popular Reading. With numerous Illustrations. By Mrs. F. FENWICK MILLER. Small 8vo, cloth limp, 2s. 6d.

Milton (J. L.), Works by:
The Hygiene of the Skin. A Concise Set of Rules for the Management of the Skin; with Directions for Diet, Wines, Soaps, Baths, &c. Small 8vo, 1s.; cloth extra, 1s. 6d

The Bath in Diseases of the Skin. Small 8vo, 1s.; cloth extra, 1s. 6d.

The Laws of Life, and their Relation to Diseases of the Skin. Small 8vo, 1s.; cloth extra, 1s. 6d.

Moncrieff. — **The Abdication ;** or, Time Tries All. An Historical Drama. By W. D. Scott-Moncrieff. With Seven Etchings by John Pettie, R.A., W. Q. Orchardson, R.A., J. MacWhirter, A.R.A., Colin Hunter, R. Macbeth, and Tom Graham. Large 4to, bound in buckram, 21s.

Murray (D. Christie), Novels by. Crown 8vo, cloth extra, 3s. 6d. each ; post 8vo, illustrated boards, 2s. each.

A Life's Atonement.
A Model Father.
Joseph's Coat.
Coals of Fire.
By the Gate of the Sea.

Crown 8vo, cloth extra, 3s. 6d. each.

Val Strange: A Story of the Primrose Way.
Hearts.
The Way of the World. Three Vols., crown 8vo.

North Italian Folk. By Mrs. Comyns Carr. Illust. by Randolph Caldecott. Square 8vo, cloth extra, 7s. 6d.

Number Nip (Stories about), the Spirit of the Giant Mountains. Retold for Children by Walter Grahame. With Illustrations by J. Moyr Smith. Post 8vo, cloth extra, 5s.

Nursery Hints: A Mother's Guide in Health and Disease. By N. E. Davies, L.R.C.P. Crown 8vo, 1s. ; cloth, 1s. 6d.

Oliphant. — **Whiteladies: A** Novel. With Illustrations by Arthur Hopkins and Henry Woods. Crown 8vo, cloth extra, 3s. 6d. ; post 8vo, illustrated boards, 2s.

O'Reilly.—Phœbe's Fortunes : A Novel. With Illustrations by Henry Tuck. Post 8vo, illustrated boards, 2s.

O'Shaughnessy (Arth.), Works by :
Songs of a Worker. Fcap. 8vo, cloth extra, 7s. 6d.
Music and Moonlight. Fcap. 8vo, cloth extra, 7s. 6d.
Lays of France. Crown 8vo, cloth extra, 10s. 6d.

Ouida, Novels by. Crown 8vo, cloth extra, 5s. each ; post 8vo, illustrated boards, 2s. each.

Held in Bondage.	A Dog of Flanders.
Strathmore.	Pascarel.
Chandos.	Signa.
Under Two Flags.	In a Winter City.
Cecil Castle-	Ariadne.
maine's Gage.	Friendship.
Idalia.	Moths.
Tricotrin.	Pipistrello.
Puck.	A Village Com-
Folle Farine.	mune.
TwoLittleWooden	Bimbi.
Shoes.	In Maremma.

Wanda: A Novel. Crown 8vo, cloth extra, 5s.

Frescoes: Dramatic Sketches. Crown 8vo, cloth extra, 10s. 6d.

Bimbi: Presentation Edition. Sq. 8vo, cloth gilt, cinnamon edges, 7s. 6d.

Princess Napraxine. Three Vols., crown 8vo. [Shortly.

Wisdom, Wit, and Pathos. Selected from the Works of Ouida by F. Sydney Morris. Small crown 8vo, cloth extra, 5s.

Page (H. A.), Works by :
Thoreau: His Life and Aims: A Study. With a Portrait. Post 8vo, cloth limp, 2s. 6d.
Lights on the Way: Some Tales within a Tale. By the late J. H. Alexander, B.A. Edited by H. A. Page. Crown 8vo, cloth extra, 6s.

Pascal's Provincial Letters. A New Translation, with Historical Introduction and Notes, by T. M'Crie, D.D. Post 8vo, cloth limp, 2s.

Paul Ferroll :
Post 8vo, illustrated boards, 2s. each
Paul Ferroll : A Novel.
Why Paul Ferroll Killed His Wife.

Paul.—Gentle and Simple. By Margaret Agnes Paul. With a Frontispiece by Helen Paterson. Cr. 8vo, cloth extra, 3s. 6d. ; post 8vo, illustrated boards, 2s.

Payn (James), Novels by.
Crown 8vo, cloth extra, 3s. 6d. each ;
post 8vo, illustrated boards, 2s. each.

Lost Sir Massingberd.
The Best of Husbands
Walter's Word.
Halves. | Fallen Fortunes.
What He Cost Her.
Less Black than We're Painted.
By Proxy. | High Spirits.
Under One Roof. | Carlyon's Year.
A Confidential Agent.
Some Private Views.
From Exile.
A Grape from   Thorn.
For Cash Only.

Post 8vo, illustrated boards, 2s. each.

A Perfect Treasure.
Bentinck's Tutor.
Murphy's Master.
A County Family. | At Her Mercy.
A Woman's Vengeance.
Cecil's Tryst.
The Clyffards of Clyffe.
The Family Scapegrace
The Foster Brothers.
Found Dead.
Gwendoline's Harvest.
Humorous Stories.
Like Father, Like Son.
A Marine Residence.
Married Beneath Him.
Ł Abbey.
Not Wooed, but Won.
Two Hundred Pounds Reward.

Kit: A Memory. Crown 8vo, cloth
extra, 3s. 6d.

The Canon's Ward.   Three Vols.,
crown 8vo.

Pennell (H. Cholmondeley),
Works by : Post 8vo, cloth limp,
2s. 6d. each.

Puck on Pegasus. With Illustrations.
The Muses of Mayfair. Vers de
Société, Selected and Edited by H.
C. PENNELL.
Pegasus Re-Saddled. With Ten full-
page Illusts. by G. DU MAURIER,

Phelps.—Beyond the Gates.
By ELIZABETH STUART PHELPS,
Author of "The Gates Ajar." Crown
8vo, cloth extra, 2s. 6d. *Published by
special arrangement with the Author,
and Copyright in England and its
Dependencies.*

Pirkis.—Trooping with Crows :
A Story. By CATHERINE PIRKIS. Fcap.
8vo, picture cover, 1s.

Planche (J. R.), Works by :
The Cyclopœdia of Costume ; or,
A Dictionary of Dress—Regal, Ec-
clesiastical, Civil, and Military—from
the Earliest Period in England to the
Reign of George the Third. Including Notices of Contemporaneous
. Fashions on the Continent, and a
General History of the Costumes of
the Principal Countries of Europe.
Two Vols., demy 4to, half morocco,
profusely Illustrated with Coloured
and Plain Plates and Woodcuts,
£7 7s. The Vols. may also be had
separately (each complete in itself)
at £3 13s. 6d. each : Vol. I. THE
DICTIONARY. Vol. II. A GENERAL
HISTORY OF COSTUME IN EUROPE.

The Pursuivant of Arms ; or, Heraldry Founded upon Facts. With
Coloured Frontispiece and 200 Illustrations. Crown 8vo, cloth extra,
7s. 6d.

Songs and Poems, from 1819 to 1879.
Edited, with an Introduction, by his
Daughter, Mrs. MACKARNESS. Crown
8vo, cloth extra, 6s.

Play-time : Sayings and Doings
of Babyland. By EDWARD STANFORD.
Large 4to, handsomely printed in
Colours, 5s.

Plutarch's Lives of Illustrious
Men. Translated from the Greek,
with Notes Critical and Historical, and
a Life of Plutarch, by JOHN and
WILLIAM LANGHORNE. Two Vols.,
8vo, cloth extra, with Portraits, 10s. 6d.

Poe (Edgar Allan) :—
The Choice Works, in Prose and
Poetry, of EDGAR ALLAN POE. With
an Introductory Essay by CHARLES
BAUDELAIRE, Portrait and Facsimiles. Crown 8vo, cloth extra,
7s. 6d.

The Mystery of Marie Roget, and
other Stories. Post 8vo, illustrated
boards, 2s.

Pope's Poetical Works. Com-
plete in One Volume. Post 8vo, cloth
limp, 2s.

Price (E. C.), Novels by :
Valentina : A Sketch. With a Frontispiece by HAL LUDLOW. Crown
8vo, cloth extra, 3s. 6d. ; post 8vo,
illustrated boards, 2s.

The Foreigners. Crown 8vo, cloth
extra, 3s. 6d.   *[Shortly*

Proctor (Richd. A.), Works by :

Flowers of the Sky. With 55 Illustrations. Small crown 8vo, cloth extra, **4s. 6d.**

Easy Star Lessons. With Star Maps for Every Night in the Year, Drawings of the Constellations, &c. Crown 8vo, cloth extra, **6s.**

Familiar Science Studies. Crown 8vo, cloth extra, **7s. 6d.**

Rough Ways made Smooth : A Series of Familiar Essays on Scientific Subjects. Cr. 8vo, cloth extra, **6s.**

Our Place among Infinities : A Series of Essays contrasting our Little Abode in Space and Time with the Infinities Around us. Crown 8vo, cloth extra, **6s.**

The Expanse of Heaven : A Series of Essays on the Wonders of the Firmament. Cr. 8vo, cloth extra, **6s.**

Saturn and its System. New and Revised Edition, with 13 Steel Plates. Demy 8vo, cloth extra, **10s. 6d.**

The Great Pyramid : Observatory, Tomb, and Temple. With Illustrations. Crown 8vo, cloth extra, **6s.**

Mysteries of Time and Space. With Illustrations. Crown 8vo, cloth extra, **7s. 6d.**

Wages and Wants of Science Workers. Crown 8vo, **1s 6d.**

Pyrotechnist's Treasury (The);

or, Complete Art of Making Fireworks. By THOMAS KENTISH. With numerous Illustrations. Cr. 8vo, cl. extra, **4s. 6d.**

Rabelais' Works. Faithfully

Translated from the French, with variorum Notes, and numerous characteristic Illustrations by GUSTAVE DORÉ. Crown 8vo, cloth extra, **7s. 6d.**

Rambosson.—Popular Astro-

nomy. By J. RAMBOSSON, Laureate of the Institute of France. Translated by C. B. PITMAN. Crown 8vo, cloth gilt, with numerous Illustrations, and a beautifully executed Chart of Spectra, **7s. 6d.**

Reader's Handbook (The) of

Allusions, References, Plots, and Stories. By the Rev. Dr. BREWER. Third Edition, revised throughout, with a New Appendix, containing a COMPLETE ENGLISH BIBLIOGRAPHY. Crown 8vo, 1,400 pages, cloth extra, **7s. 6d.**

Richardson. — A Ministry of

Health, and other Papers. By BENJAMIN WARD RICHARDSON, M.D., &c. Crown 8vo, cloth extra, **6s.**

Reade (Charles, D.C.L.), Novels

by. Post 8vo, illustrated boards, **2s.** each ; or crown 8vo, cloth extra, Illustrated, **3s. 6d.** each.

Peg Woffington. Illustrated by S. L. FILDES, A.R.A.

Christie Johnstone. Illustrated by WILLIAM SMALL.

It is Never Too Late to Mend. Illustrated by G. J. PINWELL.

The Course of True Love Never did run Smooth. Illustrated by HELEN PATERSON.

The Autobiography of a Thief ; Jack of all Trades; and James Lambert. Illustrated by MATT STRETCH.

Love me Little, Love me Long. Illustrated by M. ELLEN EDWARDS.

The Double Marriage. Illustrated by Sir JOHN GILBERT, R.A., and CHARLES KEENE.

The Cloister and the Hearth. Illustrated by CHARLES KEENE.

Hard Cash. Illustrated by F. W. LAWSON.

Griffith Gaunt. Illustrated by S. L. FILDES, A.R.A., and WM. SMALL.

Foul Play. Illustrated by GEORGE DU MAURIER.

Put Yourself in His Place. Illustrated by ROBERT BARNES.

A Terrible Temptation. Illustrated by EDW. HUGHES and A. W. COOPER.

The Wandering Heir. Illustrated by HELEN PATERSON, S. L. FILDES, A.R.A., CHARLES GREEN, and HENRY WOODS, A.R.A.

A Simpleton. Illustrated by KATE CRAUFORD.

A Woman-Hater. Illustrated by THOS. COULDERY.

Readiana. With a Steel Plate Portrait of CHARLES READE.

A New Collection of Stories. In Three Vols., crown 8vo. [*Preparing.*

Riddell (Mrs. J. H.), Novels by :

Crown 8vo, cloth extra, **3s. 6d.** each ; post 8vo, illustrated boards, **2s.** each.

Her Mother's Darling.

The Prince of Wales's Garden Party.

Rimmer (Alfred), Works by :

Our Old Country Towns. With over 50 Illusts. Sq. 8vo, cloth gilt, **10s. 6d.**

Rambles Round Eton and Harrow. 50 Illusts. Sq. 8vo, cloth gilt, **10s. 6d.**

About England with Dickens. With 58 Illustrations by ALFRED RIMMER and C. A. VANDERHOOF. Square 8vo, cloth gilt, **10s. 6d.**

Robinson (F. W.), Novels by:
Women are Strange. Cr. 8vo, cloth extra, 3s. 6d.; post 8vo, illust. bds., 2s.
The Hands of Justice. Crown 8vo, cloth extra, 3s. 6d.

Robinson (Phil), Works by:
The Poets' Birds. Crown 8vo, cloth extra, 7s. 6d.
The Poets' Beasts. Crown 8vo, cloth extra, 7s. 6d. [*In preparation.*]

Robinson Crusoe: A beautiful reproduction of Major's Edition, with 37 Woodcuts and Two Steel Plates by GEORGE CRUIKSHANK, choicely printed. Crown 8vo, cloth extra, 7s. 6d. 100 Large-Paper copies, printed on hand-made paper, with India proofs of the Illustrations, price 36s.

Rochefoucauld's Maxims and Moral Reflections. With Notes, and an Introductory Essay by SAINTE-BEUVE. Post 8vo, cloth limp, 2s.

Roll of Battle Abbey, The; or, A List of the Principal Warriors who came over from Normandy with William the Conqueror, and Settled in this Country, A.D. 1066-7. With the principal Arms emblazoned in Gold and Colours. Handsomely printed, price 5s.

Rowley (Hon. Hugh), Works by:
Post 8vo, cloth limp, 2s. 6d. each.
Puniana: Riddles and Jokes. With numerous Illustrations.
More Puniana. Profusely Illustrated.

Russell (Clark).—Round the Galley-Fire. By W. CLARK RUSSELL, Author of "The Wreck of the *Grosvenor.*" Cr. 8vo, cloth extra, 6s.

Sala.—Gaslight and Daylight. By GEORGE AUGUSTUS SALA. Post 8vo, illustrated boards, 2s.

Sanson.—Seven Generations of Executioners: Memoirs of the Sanson Family (1688 to 1847). Edited by HENRY SANSON. Crown 8vo, cloth extra, 3s. 6d.

Saunders (John), Novels by:
Crown 8vo, cloth extra, 3s. 6d. each; post 8vo, illustrated boards, 2s. each.
Bound to the Wheel.
One Against the World.
Guy Waterman.
The Lion in the Path.
The Two Dreamers.

Science Gossip: An Illustrated Medium of Interchange for Students and Lovers of Nature. Edited by J. E. TAYLOR, F.L.S., &c. Devoted to Geology, Botany, Physiology, Chemistry, Zoology, Microscopy, Telescope, Physiography, &c. Price 4d. Monthly; or 5s. per year, post free. It contains Original Illustrated Articles by the best-known Writers and Workers of the day. A Monthly Summary of Discovery and Progress in every department of Natural Science is given. Large space is devoted to Scientific "Notes and Queries," thus enabling every lover of nature to chronicle his own original observations, or get his special difficulties settled. For active workers and collectors the "Exchange Column" has long proved a well and widely known means of barter and exchange. The column devoted to "Answers to Correspondents" has been found helpful to students requiring personal help in naming specimens, &c. The Volumes of *Science Gossip* for the last eighteen years contain an unbroken History of the advancement of Natural Science within that period. Each Number contains a Coloured Plate and numerous Woodcuts. Vols. I. to XIV. may be had at 7s. 6d. each; and Vols. XV. to XIX. (1883), at 5s. each.

"Secret Out" Series, The: Crown 8vo, cloth extra, profusely Illustrated, 4s. 6d. each.
The Secret Out: One Thousand Tricks with Cards, and other Recreations; with Entertaining Experiments in Drawing-room or "White Magic." By W. H. CREMER. 300 Engravings.
The Pyrotechnist's Treasury; or, Complete Art of Making Fireworks. By THOMAS KENTISH. With numerous Illustrations.
The Art of Amusing: A Collection of Graceful Arts, Games, Tricks, Puzzles, and Charades. By FRANK BELLEW. With 300 Illustrations.
Hanky-Panky: Very Easy Tricks, Very Difficult Tricks, White Magic, Sleight of Hand. Edited by W. H. CREMER. With 200 Illustrations.
The Merry Circle: A Book of New Intellectual Games and Amusements. By CLARA BELLEW. With many Illustrations.
Magician's Own Book: Performances with Cups and Balls, Eggs, Hats, Handkerchiefs, &c. All from actual Experience. Edited by W. H. CREMER. 200 Illustrations.

THE "SECRET OUT" SERIES, *continued—*
Magic No Mystery: Tricks with Cards, Dice, Balls, &c., with fully descriptive Directions; the Art of Secret Writing; Training of Performing Animals, &c. With Coloured Frontispiece and many Illustrations.

Senior (William), Works by :

Travel and Trout In the Antipodes. Crown 8vo, cloth extra, 6s.

By Stream and Sea. Post 8vo, cloth limp, 2s. 6d.

Seven Sagas (The) of Prehistoric Man. By JAMES H. STODDART, Author of "The Village Life." Crown 8vo, cloth extra, 6s.

Shakespeare :

The First Folio Shakespeare.—MR. WILLIAM SHAKESPEARE's Comedies, Histories, and Tragedies. Published according to the true Originall Copies. London, Printed by ISAAC IAGGARD and ED. BLOUNT. 1623.—A Reproduction of the extremely rare original, in reduced facsimile, by a photographic process—ensuring the strictest accuracy in every detail. Small 8vo, half-Roxburghe, 7s. 6d.

The Lansdowne Shakespeare. Beautifully printed in red and black, in small but very clear type. With engraved facsimile of DROESHOUT's Portrait. Post 8vo, cloth extra, 7s. 6d.

Shakespeare for Children: Tales from Shakespeare. By CHARLES and MARY LAMB. With numerous Illustrations, coloured and plain, by J. MOYR SMITH. Crown 4to, cloth gilt, 6s.

The Handbook of Shakespeare Music. Being an Account of 350 Pieces of Music, set to Words taken from the Plays and Poems of Shakespeare, the compositions ranging from the Elizabethan Age to the Present Time. By ALFRED ROFFE. 4to, half-Roxburghe, 7s.

A Study of Shakespeare. By ALGERNON CHARLES SWINBURNE. Crown 8vo, cloth extra, 8s.

Shelley's Complete Works, in Four Vols., post 8vo, cloth limp, 8s. ; or separately, 2s. each. Vol. I. contains his Early Poems, Queen Mab, &c., with an Introduction by LEIGH HUNT; Vol. II., his Later Poems, Laon and Cythna, &c.; Vol. III., Posthumous Poems, the Shelley Papers, &c. : Vol. IV., his Prose Works, including A Refutation of Deism, Zastrozzi, St. Irvyne, &c.

Sheridan's Complete Works, with Life and Anecdotes. Including his Dramatic Writings, printed from the Original Editions, his Works in Prose and Poetry, Translations, Speeches, Jokes, Puns, &c. With a Collection of Sheridaniana. Crown 8vo, cloth extra, gilt, with 10 full-page Tinted Illustrations, 7s. 6d.

Short Sayings of Great Men. With Historical and Explanatory Notes by SAMUEL A. BENT, M.A. Demy 8vo, cloth extra, 7s. 6d.

Sidney's (Sir Philip) Complete Poetical Works, including all those in "Arcadia." With Portrait, Memorial-Introduction, Essay on the Poetry of Sidney, and Notes, by the Rev. A. B. GROSART, D.D. Three Vols., crown 8vo, cloth boards, 18s.

Signboards: Their History. With Anecdotes of Famous Taverns and Remarkable Characters. By JACOB LARWOOD and JOHN CAMDEN HOTTEN. Crown 8vo, cloth extra, with 100 Illustrations, 7s. 6d.

Sims (G. R.), Works by :

How the Poor Live. With 60 Illustrations by FRED. BARNARD. Large 4to, 1s.

Horrible London. Reprinted, with Additions, from the *Daily News*. Large 4to, 6d.   [*Shortly.*

Sketchley.—A Match in the Dark. By ARTHUR SKETCHLEY. Post 8vo, illustrated boards, 2s.

Slang Dictionary, The: Etymological, Historical, and Anecdotal. Crown 8vo, cloth extra, gilt, 6s. 6d.

Smith (J. Moyr), Works by :

The Prince of Argolis: A Story of the Old Greek Fairy Time. By J. MOYR SMITH. Small 8vo, cloth extra, with 130 Illustrations, 3s. 6d.

Tales of Old Thule. Collected and Illustrated by J. MOYR SMITH. Crown 8vo, cloth gilt, profusely Illustrated, 6s.

The Wooing of the Water Witch: A Northern Oddity. By EVAN DALDORNE. Illustrated by J. MOYR SMITH. Small 8vo, cloth extra, 6s.

South-West, The New : Travelling Sketches from Kansas, New Mexico, Arizona, and Northern Mexico. By ERNST VON HESSE-WARTEGG. With 100 fine Illustrations and 3 Maps. 8vo, cloth extra, 14s. [*In preparation.*

Spalding.—Elizabethan Demonology: An Essay in Illustration of the Belief in the Existence of Devils, and the Powers possessed by Them. By T. ALFRED SPALDING, LL.B. Crown 8vo, cloth extra, 5s.

Speight. — The Mysteries of Heron Dyke. By T. W. SPEIGHT. With a Frontispiece by M. ELLEN EDWARDS. Crown 8vo, cloth extra, 3s. 6d. ; post 8vo, illustrated boards, 2s.

Spenser for Children. By M. H. TOWRY. With Illustrations by WALTER J. MORGAN. Crown 4to, with Coloured Illustrations, cloth gilt, 6s.

Staunton.—Laws and Practice of Chess; Together with an Analysis of the Openings, and a Treatise on End Games. By HOWARD STAUNTON. Edited by ROBERT B. WORMALD. New Edition, small cr. 8vo, cloth extra, 5s.

Sterndale.—The Afghan Knife: A Novel. By ROBERT ARMITAGE STERNDALE. Cr. 8vo, cloth extra, 3s. 6d.; post 8vo, illustrated boards, 2s.

Stevenson (R.Louis),Works by :

Travels with a Donkey in the Cevennes. Frontispiece by WALTER CRANE. Post 8vo, cloth limp, 2s. 6d.

An Inland Voyage. With a Frontispiece by WALTER CRANE. Post 8vo, cloth limp, 2s. 6d.

Virginibus Puerisque, and other Papers. Crown 8vo, cloth extra, 6s.

Familiar Studies of Men and Books. Crown 8vo, cloth extra, 6s.

New Arabian Nights. Crown 8vo, cl. extra, 6s.; post 8vo, illust. bds., 2s.

The Silverado Squatters. With Frontispiece. Cr. 8vo, cloth extra, 6s.

St.John.—A Levantine Family. By BAYLE ST. JOHN. Post 8vo, illustrated boards, 2s.

Stoddard.—Summer Cruising in the South Seas. By CHARLES WARREN STODDARD. Illustrated by WALLIS MACKAY. Crown 8vo, cloth extra, 3s. 6d.

St. Pierre.—Paul and Virginia, and The Indian Cottage. By BERNARDIN DE ST. PIERRE. Edited, with Life, by the Rev. E. CLARKE. Post 8vo, cloth limp, 2s.

Stories from Foreign Novelists. With Notices of their Lives and Writings. By HELEN and ALICE ZIMMERN ; and a Frontispiece. Crown 8vo cloth extra, 3s. 6d.

Strutt's Sports and Pastimes of the People of England; including the Rural and Domestic Recreations, May Games, Mummeries, Shows, Processions, Pageants, and Pompous Spectacles, from the Earliest Period to the Present Time. With 140 Illustrations. Edited by WILLIAM HONE. Crown 8vo, cloth extra, 7s. 6d.

Suburban Homes (The) of London: A ˙ Residential Guide to Favourite London Localities, their Society, Celebrities, and Associations. With Notes on their Rental, Rates, and House Accommodation. With a Map of Suburban London. Crown 8vo, cloth extra, 7s. 6d.

Swift's Choice Works, in Prose and Verse. With Memoir, Portrait, and Facsimiles of the Maps in the Original Edition of " Gulliver's Travels." Cr. 8vo, cloth extra, 7s. 6d.

Swinburne (Algernon C.), Works by :

The Queen Mother and Rosamond. Fcap. 8vo, 5s.

Atalanta in Calydon. Crown 8vo, 6s.

Chastelard. A Tragedy. Crown 8vo, 7s.

Poems and Ballads. FIRST SERIES. Fcap. 8vo, 9s. Also in crown 8vo, at same price.

Poems and Ballads. SECOND SERIES. Fcap. 8vo, 9s. Also in crown 8vo, at same price.

Notes on Poems and Reviews. 8vo, 1s.

William Blake: A Critical Essay. With Facsimile Paintings. Demy 8vo, 16s.

Songs before Sunrise. Crown 8vo, 10s. 6d.

Bothwell: A Tragedy. Crown 8vo, 12s. 6d.

George Chapman: An Essay. Crown 8vo, 7s.

Songs of Two Nations. Cr. 8vo, 6s.

Essays and Studies. Crown 8vo, 12s.

Erechtheus: A Tragedy. Crown 8vo, 6s.

Note of an English Republican on the Muscovite Crusade. 8vo, 1s.

A Note on Charlotte Bronte. Crown 8vo, 6s.

A Study of Shakespeare. Crown 8vo, 8s.

Songs of the Springtides. Crown 8vo, 6s.

Studies in Song. Crown 8vo, 7s.

A. C. Swinburne's Works, *continued—*

Mary Stuart: A Tragedy. Crown 8vo, 8s.

Tristram of Lyonesse, and other Poems. Crown 8vo, 9s.

A Century of Roundels. Small 4to, cloth extra, 8s.

Syntax's (Dr.) Three Tours: In Search of the Picturesque, in Search of Consolation, and in Search of a Wife. With the whole of Rowlandson's droll page Illustrations in Colours and a Life of the Author by J. C. Hotten. Medium 8vo, cl. extra, 7s. 6d.

Taine's History of English Literature. Translated by Henry Van Laun. Four Vols., small 8vo, cloth boards, 30s.—Popular Edition, Two Vols., crown 8vo, cloth extra, 15s.

Taylor (Dr.).—The Sagacity and Morality of Plants: A Sketch of the Life and Conduct of the Vegetable Kingdom. By J. E. Taylor, F.L.S., &c. With Coloured Frontispiece and 100 Illustrations. Crown 8vo, cloth extra, 7s. 6d.

Taylor's (Bayard) Diversions of the Echo Club: Burlesques of Modern Writers. Post 8vo, cl. limp, 2s.

Taylor's (Tom) Historical Dramas: "Clancarty," "Jeanne Darc," "'Twixt Axe and Crown," "The Fool's Revenge," "Arkwright's Wife," "Anne Boleyn," "Plot and Passion." One Vol., crown 8vo, cloth extra, 7s. 6d.

*** The Plays may also be had separately, at 1s. each.

Thackerayana: Notes and Anecdotes. Illustrated by Hundreds of Sketches by William Makepeace Thackeray, depicting Humorous Incidents in his School-life, and Favourite Characters in the books of his every-day reading. With Coloured Frontispiece. Cr. 8vo, cl. extra, 7s. 6d.

Thomas (Bertha), Novels by. Crown 8vo, cloth extra, 3s. 6d. each; post 8vo, illustrated boards, 2s. each.

Cressida.
Proud Maisie.
The Violin-Player.

Thomson's Seasons and Castle of Indolence. With a Biographical and Critical Introduction by Allan Cunningham, and over 50 fine Illustrations on Steel and Wood. Crown 8vo, cloth extra, gilt edges, 7s. 6d.

Thomas (M.).—A Fight for Life: A Novel. By W. Moy Thomas. Post 8vo, illustrated boards, 2s.

Thornbury (Walter), Works by

Haunted London. Edited by Edward Walford, M.A. With Illustrations by F. W. Fairholt, F.S.A. Crown 8vo, cloth extra, 7s. 6d.

The Life and Correspondence of J. M. W. Turner. Founded upon Letters and Papers furnished by his Friends and fellow Academicians. With numerous Illustrations in Colours, facsimiled from Turner's Original Drawings. Crown 8vo, cloth extra, 7s. 6d.

Old Stories Re-told. Post 8vo, cloth limp, 2s. 6d.

Tales for the Marines. Post 8vo, illustrated boards, 2s.

Timbs (John), Works by:

The History of Clubs and Club Life in London. With Anecdotes of its Famous Coffee-houses, Hostelries, and Taverns. With numerous Illustrations. Cr. 8vo, cloth extra, 7s. 6d.

English Eccentrics and Eccentricities: Stories of Wealth and Fashion, Delusions, Impostures, and Fanatic Missions, Strange Sights and Sporting Scenes, Eccentric Artists, Theatrical Folks, Men of Letters, &c. With nearly 50 Illusts. Crown 8vo, cloth extra, 7s. 6d.

Torrens. — The Marquess Wellesley, Architect of Empire. An Historic Portrait. By W. M. Torrens, M.P. Demy 8vo, cloth extra, 14s.

Trollope (Anthony), Novels by: Crown 8vo, cloth extra, 3s. 6d. each; post 8vo, illustrated boards, 2s. each.

The Way We Live Now.
The American Senator.
Kept in the Dark.
Frau Frohmann.
Marion Fay.

Mr. Scarborough's Family. Crown 8vo, cloth extra, 3s. 6d.

The Land-Leaguers. Crown 8vo, cloth extra, 3s. 6d. [*Shortly.*

Trollope (Frances E.), Novels by

Like Ships upon the Sea. Crown 8vo, cloth extra, 3s. 6d.; post 8vo, illustrated boards, 2s.

Mabel's Progress. Crown 8vo, cloth extra, 3s. 6d.

Anne Furness. Crown 8vo, cloth extra, 3s. 6d.

Trollope (T. A.).—Diamond Cut
Diamond, and other Stories. By
THOMAS ADOLPHUS TROLLOPE. Crown
8vo, cloth extra, 3s. 6d.; post 8vo,
illustrated boards, 2s.

Tytler (Sarah), Novels by:
Crown 8vo, cloth extra, 3s. 6d. each;
post 8vo, illustrated boards, 2s. each.
What She Came Through.
The Bride's Pass.

Van Laun.—History of French
Literature. By HENRY VAN LAUN.
Complete in Three Vols., demy 8vo,
cloth boards, 7s. 6d. each.

Villari.—A Double Bond: A
Story. By LINDA VILLARI. Fcap.
8vo, picture cover, 1s.

Walcott.— Church Work and
Life in English Minsters; and the
English Student's Monasticon. By the
Rev. MACKENZIE E. C. WALCOTT, B.D.
Two Vols., crown 8vo, cloth extra,
with Map and Ground-Plans, 14s.

Walford (Edw., M.A.),Works by:
The County Families of the United
Kingdom. Containing Notices of
the Descent, Birth, Marriage, Educa-
tion, &c., of more than 12,000 dis-
tinguished Heads of Families, their
Heirs Apparent or Presumptive, the
Offices they hold or have held, their
Town and Country Addresses, Clubs,
&c. Twenty-fourth Annual Edition,
for 1884, cloth, full gilt, 50s. [Shortly.
The Shilling Peerage (1884). Con-
taining an Alphabetical List of the
House of Lords, Dates of Creation,
Lists of Scotch and Irish Peers,
Addresses, &c. 32mo, cloth, 1s.
Published annually.
The Shilling Baronetage (1884).
Containing an Alphabetical List of
the Baronets of the United Kingdom,
short Biographical Notices, Dates
of Creation, Addresses, &c. 32mo,
cloth, 1s. Published annually.
The Shilling Knightage (1884). Containing an Alphabetical List of the
Knights of the United Kingdom,
short Biographical Notices, Dates
of Creation, Addresses, &c. 32mo,
cloth, 1s. Published annually.
The Shilling House of Commons
(1884). Containing a List of all the
Members of the British Parliament,
their Town and Country Addresses,
&c. 32mo, cloth, 1s. Published
annually.

EDW. WALFORD'S WORKS, continued—
The Complete Peerage, Baronetage, Knightage, and House of
Commons (1884). In One Volume,
royal 32mo, cloth extra, gilt edges,
5s. Published annually.
Haunted London. By WALTER
THORNBURY. Edited by EDWARD
WALFORD, M.A. With Illustrations
by F. W. FAIRHOLT, F.S.A. Crown
8vo, cloth extra, 7s. 6d.

Walton and Cotton's Complete
Angler; or, The Contemplative Man's
Recreation; being a Discourse of
Rivers, Fishponds, Fish and Fishing,
written by IZAAK WALTON; and In-
structions how to Angle for a Trout or
Grayling in a clear Stream, by CHARLES
COTTON. With Original Memoirs and
Notes by Sir HARRIS NICOLAS, and
61 Copperplate Illustrations. Large
crown 8vo, cloth antique, 7s. 6d.

Wanderer's Library, The:
Crown 8vo, cloth extra, 3s. 6d. each.
Wanderings in Patagonia; or, Life
among the Ostrich Hunters. By
JULIUS BEERBOHM. Illustrated.
Camp Notes: Stories of Sport and
Adventure in Asia, Africa, and
America. By FREDERICK BOYLE.
Savage Life. By FREDERICK BOYLE.
Merrie England in the Olden Time.
By GEORGE DANIEL. With Illustra-
tions by ROBT. CRUIKSHANK.
Circus Life and Circus Celebrities
By THOMAS FROST.
The Lives of the Conjurers. By
THOMAS FROST.
The Old Showmen and the Old
London Fairs. By THOMAS FROST.
Low-Life Deeps. An Account of the
Strange Fish to be found there. By
JAMES GREENWOOD.
The Wilds of London. By JAMES
GREENWOOD.
Tunis: The Land and the People.
By the Chevalier de HESSE-WAR-
TEGG. With 22 Illustrations.
The Life and Adventures of a Cheap
Jack. By One of the Fraternity.
Edited by CHARLES HINDLEY.
The World Behind the Scenes. By
PERCY FITZGERALD.
Tavern Anecdotes and Sayings
Including the Origin of Signs, and
Reminiscences connected with Ta-
verns, Coffee Houses, Clubs, &c.
By CHARLES HINDLEY. With Illusts.
The Genial Showman: Life and Ad-
ventures of Artemus Ward. By E. P
HINGSTON. With a Frontispiece.

THE WANDERER'S LIBRARY, *continued*—

The Story of the London Parks. By JACOB LARWOOD. With Illustrations.

London Characters. By HENRY MAYHEW. Illustrated.

Seven Generations of Executioners: Memoirs of the Sanson Family (1688 to 1847). Edited by HENRY SANSON.

Summer Cruising in the South Seas. By CHARLES WARREN STODDARD. Illustrated by WALLIS MACKAY.

Warner.—A Roundabout Journey. By CHARLES DUDLEY WARNER, Author of "My Summer in a Garden." Crown 8vo, cloth extra, 6s.

Warrants, &c. :—

Warrant to Execute Charles I. An exact Facsimile, with the Fifty-nine Signatures, and corresponding Seals. Carefully printed on paper to imitate the Original, 22 in. by 14 in. Price 2s.

Warrant to Execute Mary Queen of Scots. An exact Facsimile, including the Signature of Queen Elizabeth, and a Facsimile of the Great Seal. Beautifully printed on paper to imitate the Original MS. Price 2s.

Magna Charta. An Exact Facsimile of the Original Document in the British Museum, printed on fine plate paper, nearly 3 feet long by 2 feet wide, with the Arms and Seals emblazoned in Gold and Colours. Price 5s.

The Roll of Battle Abbey; or, A List of the Principal Warriors who came over from Normandy with William the Conqueror, and Settled in this Country, A.D. 1066-7. With the principal Arms emblazoned in Gold and Colours. Price 5s.

Westropp.—Handbook of Pottery and Porcelain; or, History of those Arts from the Earliest Period. By HODDER M. WESTROPP. With numerous Illustrations, and a List of Marks. Crown 8vo, cloth limp, 4s. 6d.

Whistler v. Ruskin: Art and Art Critics. By J. A. MACNEILL WHISTLER. Seventh Edition, square 8vo, 1s.

White's Natural History of Selborne. Edited, with Additions, by THOMAS BROWN, F.L.S. Post 8vo, cloth limp, 2s.

Williams (W. Mattieu, F.R.A.S.), Works by:

Science Notes. See the GENTLEMAN'S MAGAZINE. 1s. Monthly.

Science in Short Chapters. Crown 8vo, cloth extra, 7s. 6d.

A Simple Treatise on Heat. Crown 8vo, cloth limp, with Illusts., 2s. 6d.

Wilson (Dr. Andrew, F.R.S.E.), Works by:

Chapters on Evolution: A Popular History of the Darwinian and Allied Theories of Development. Second Edition. Crown 8vo, cloth extra, with 259 Illustrations, 7s. 6d.

Leaves from a Naturalist's Notebook. Post 8vo, cloth limp, 2s. 6d.

Leisure-Time Studies, chiefly Biological. Second Edition. Crown 8vo, cloth extra, with Illustrations, 6s.

Wilson (C.E.).—Persian Wit and Humour: Being the Sixth Book of the Baharistan of Jami, Translated for the first time from the Original Persian into English Prose and Verse. With Notes by C. E. WILSON, M.R.A.S., Assistant Librarian Royal Academy of Arts. Cr. 8vo, parchment binding, 4s.

Winter (J. S.), Stories by: Crown 8vo, cloth extra, 3s. 6d. each post 8vo, illustrated boards, 2s. each.

Cavalry Life.

Regimental Legends.

Wood.—Sabina: A Novel. By Lady WOOD. Post 8vo, illustrated boards, 2s.

Words, Facts, and Phrases: A Dictionary of Curious, Quaint, and Out-of-the-Way Matters. By ELIEZER EDWARDS. Cr. 8vo, half-bound, 12s. 6d.

Wright (Thomas), Works by:

Caricature History of the Georges. (The House of Hanover.) With 400 Pictures, Caricatures, Squibs, Broadsides, Window Pictures, &c. Crown 8vo, cloth extra, 7s. 6d.

History of Caricature and of the Grotesque in Art, Literature, Sculpture, and Painting. Profusely Illustrated by F. W. FAIRHOLT, F.S.A. Large post 8vo, cloth extra, 7s. 6d.

Yates (Edmund), Novels by: Post 8vo, illustrated boards 2s. each.

Castaway.

The Forlorn Hope.

Land at Last.

NOVELS BY THE BEST AUTHORS.

At every Library.

Princess Napraxine. By OUIDA. Three Vols. [*Shortly.*

Dorothy Forster. By WALTER BESANT. Three Vols. [*Shortly.*

The New Abelard. By ROBERT BU-CHANAN. Three Vols.

Fancy-Free, &c. By CHARLES GIBBON. Three Vols.

Ione. E. LYNN LINTON. Three Vols.

The Way of the World. By D. CHRISTIE MURRAY. Three Vols.

Maid of Athens. By JUSTIN MCCARTHY, M.P. With 12 Illustrations by FRED. BARNARD. Three Vols.

The Canon's Ward. By JAMES PAYN. Three Vols.

A Real Queen. By R. E. FRANCILLON. Three Vols.

A New Collection of Stories by CHARLES READE. Three Vols. [*In preparation.*

THE PICCADILLY NOVELS.

Popular Stories by the Best Authors. LIBRARY EDITIONS, many Illustrated, crown 8vo, cloth extra, 3s. 6d. each.

BY MRS. ALEXANDER.
Maid, Wife, or Widow?

BY W. BESANT & JAMES RICE.
Ready-Money Mortiboy.
My Little Girl.
The Case of Mr. Lucraft.
This Son of Vulcan.
With Harp and Crown.
The Golden Butterfly.
By Celia's Arbour.
The Monks of Thelema.
'Twas in Trafalgar's Bay.
The Seamy Side.
The Ten Years' Tenant.
The Chaplain of the Fleet.

BY WALTER BESANT.
All Sorts and Conditions of Men.
The Captains' Room.

BY ROBERT BUCHANAN.
A Child of Nature.
God and the Man.
The Shadow of the Sword.
The Martyrdom of Madeline.
Love Me for Ever.

BY MRS. H. LOVETT CAMERON.
Deceivers Ever.
Juliet's Guardian.

BY MORTIMER COLLINS.
Sweet Anne Page.
Transmigration.
From Midnight to Midnight.

MORTIMER & FRANCES COLLINS.
Blacksmith and Scholar.
The Village Comedy.
You Play me False.

BY WILKIE COLLINS.

Antonina.	New Magdalen.
Basil.	The Frozen Deep.
Hide and Seek.	The Law and the
The Dead Secret.	Lady.
Queen of Hearts.	The Two Destinies
My Miscellanies.	Haunted Hotel
Woman in White.	The Fallen Leaves
The Moonstone.	Jezebel's Daughter
Man and Wife.	The Black Robe.
Poor Miss Finch.	Heart and Science
Miss or Mrs.?	

BY DUTTON COOK.
Paul Foster's Daughter

BY WILLIAM CYPLES.
Hearts of Gold.

BY JAMES DE MILLE.
A Castle in Spain.

BY J. LEITH DERWENT.
Our Lady of Tears. | Circe's Lovers.

PICCADILLY NOVELS, *continued—*

BY M. BETHAM-EDWARDS.
Felicia. | Kitty.

BY MRS. ANNIE EDWARDES.
Archie Lovell.

BY R. E. FRANCILLON.
Olympia. | Queen Cophetua.
One by One.

Prefaced by Sir BARTLE FRERE.
Pandurang Hari.

BY EDWARD GARRETT.
The Capel Girls.

BY CHARLES GIBBON.
Robin Gray.
For Lack of Gold.
In Love and War.
What will the World Say?
For the King.
In Honour Bound.
Queen of the Meadow.
In Pastures Green.
The Flower of the Forest.
A Heart's Problem.
The Braes of Yarrow.
The Golden Shaft.
Of High Degree.

BY THOMAS HARDY.
Under the Greenwood Tree.

BY JULIAN HAWTHORNE.
Garth.
Ellice Quentin.
Sebastian Strome.
Prince Saroni's Wife.
Dust.
Fortune's Fool.

BY SIR A. HELPS.
Ivan de Biron.

BY MRS. ALFRED HUNT.
Thornicroft's Model.
The Leaden Casket.
Self-Condemned.

BY JEAN INGELOW.
Fated to be Free.

BY HENRY JAMES, Jun.
Confidence.

BY HARRIETT JAY.
The Queen of Connaught.
The Dark Colleen.

BY HENRY KINGSLEY.
Number Seventeen.
Oakshott Castle.

PICCADILLY NOVELS, *continued—*

BY E. LYNN LINTON.
Patricia Kemball.
Atonement of Leam Dundas.
The World Well Lost.
Under which Lord?
With a Silken Thread.
The Rebel of the Family.
"My Love!"

BY HENRY W. LUCY.
Gideon Fleyce.

BY JUSTIN McCARTHY, M.P.
The Waterdale Neighbours.
My Enemy's Daughter.
Linley Rochford. | A Fair Saxon.
Dear Lady Disdain.
Miss Misanthrope.
Donna Quixote.
The Comet of a Season.

BY GEORGE MAC DONALD, LL.D.
Paul Faber, Surgeon.
Thomas Wingfold, Curate.

BY MRS. MACDONELL.
Quaker Cousins.

BY KATHARINE S. MACQUOID.
Lost Rose. | The Evil Eye.

BY FLORENCE MARRYAT.
Open! Sesame! | Written in Fire.

BY JEAN MIDDLEMASS.
Touch and Go.

BY D. CHRISTIE MURRAY
Life's Atonement. | Coals of Fire.
Joseph's Coat. | Val Strange.
A Model Father. | Hearts.
By the Gate of the Sea.

BY MRS. OLIPHANT.
Whiteladies.

BY MARGARET A. PAUL
Gentle and Simple.

BY JAMES PAYN.
Lost Sir Massing- | High Spirits.
  berd. | Under One Roof.
Best of Husbands | Carlyon's Year.
Fallen Fortunes. | A Confidential
Halves. | Agent.
Walter's Word. | From Exile.
What He Cost Her | A Grape from
Less Black than | Thorn.
  We're Painted. | For Cash Only.
By Proxy. | Kit: A Memory

PICCADILLY NOVELS, *continued—*
BY E. C. PRICE.
Valentina.
The Foreigners.

BY CHARLES READE, D.C.L.
It is Never Too Late to Mend.
Hard Cash. | Peg Woffington.
Christie Johnstone.
Griffith Gaunt.
The Double Marriage.
Love Me Little, Love Me Long.
Foul Play.
The Cloister and the Hearth.
The Course of True Love.
The Autobiography of a Thief.
Put Yourself In His Place.
A Terrible Temptation.
The Wandering Heir. | A Simpleton.
A Woman-Hater. | Readiana.

BY MRS. J. H. RIDDELL.
Her Mother's Darling.
Prince of Wales's Garden-Party.

BY F. W. ROBINSON.
Women are Strange.
The Hands of Justice.

BY JOHN SAUNDERS.
Bound to the Wheel.
Guy Waterman.
One Against the World.
The Lion in the Path
The Two Dreamers.

PICCADILLY NOVELS, *continued—*
BY T. W. SPEIGHT.
The Mysteries of Heron Dyke.

BY R. A. STERNDALE.
The Afghan Knife.

BY BERTHA THOMAS
Proud Maisie. | Cressida.
The Violin-Player.

BY ANTHONY TROLLOPE.
The Way we Live Now.
The American Senator.
Frau Frohmann.
Marion Fay.
Kept in the Dark.
Mr. Scarborough's Family.
The Land-Leaguers.

BY FRANCES E. TROLLOPE.
Like Ships upon the Sea.
Anne Furness.
Mabel's Progress.

BY T. A. TROLLOPE.
Diamond Cut Diamond.

By IVAN TURGENIEFF and Others.
Stories from Foreign Novelists.

BY SARAH TYTLER
What She Came Through.
The Bride's Pass.

BY J. S. WINTER.
Cavalry Life.
Regimental Legends.

CHEAP EDITIONS OF POPULAR NOVELS.
Post 8vo, illustrated boards, 2s. each.

BY EDMOND ABOUT.
The Fellah.

BY HAMILTON AÏDÉ.
Carr of Carrlyon. | Confidences.

BY MRS. ALEXANDER.
Maid, Wife, or Widow?

BY SHELSLEY BEAUCHAMP.
Grantley Grange.

BY W. BESANT & JAMES RICE.
Ready-Money Mortiboy.
With Harp and Crown.
This Son of Vulcan.
My Little Girl.
The Case of Mr. Lucraft.
The Golden Butterfly.

By BESANT AND RICE, *continued—*
By Celia's Arbour.
The Monks of Thelema.
'Twas in Trafalgar's Bay.
The Seamy Side.
The Ten Years' Tenant.
The Chaplain of the Fleet.
All Sorts and Conditions of Men.
The Captains' Room.

BY FREDERICK BOYLE.
Camp Notes. | Savage Life.

BY BRET HARTE.
An Heiress of Red Dog.
The Luck of Roaring Camp.
Californian Stories.
Gabriel Conroy. | Flip

Cheap Popular Novels, *continued*—

BY ROBERT BUCHANAN.
The Shadow of the Sword.
A Child of Nature.
God and the Man.
The Martyrdom of Madeline.
Love Me for Ever.

BY MRS. BURNETT.
Surly Tim.

BY MRS. LOVETT CAMERON.
Deceivers Ever. | Juliet's Guardian.

BY MACLAREN COBBAN.
The Cure of Souls.

BY C. ALLSTON COLLINS.
The Bar Sinister.

BY WILKIE COLLINS.
Antonina.	Miss or Mrs.?
Basil.	The New Magda-
Hide and Seek.	len.
The Dead Secret.	The Frozen Deep.
Queen of Hearts.	Law and the Lady.
My Miscellanies.	TheTwoDestinies
Woman in White.	Haunted Hotel.
The Moonstone.	The Fallen Leaves.
Man and Wife.	Jezebel'sDaughter
Poor Miss Finch.	The Black Robe.

BY MORTIMER COLLINS.
Sweet Anne Page.
Transmigration.
From Midnight to Midnight.
A Fight with Fortune.

MORTIMER & FRANCES COLLINS.
Sweet and Twenty. | Frances.
Blacksmith and Scholar.
The Village Comedy.
You Play me False.

BY DUTTON COOK.
Leo. | Paul Foster's Daughter.

BY J. LEITH DERWENT.
Ou Lady of Tears.

BY CHARLES DICKENS.
Sketches by Boz.
The Pickwick Papers.
Oliver Twist.
Nicholas Nickleby.

BY MRS. ANNIE EDWARDES.
A Point of Honour. | Archie Lovell.

BY M. BETHAM-EDWARDS.
Felicia. | Kitty.

BY EDWARD EGGLESTON.
Roxy.

Cheap Popular Novels, *continued*—

BY PERCY FITZGERALD.
Bella Donna. | Never Forgotten.
The Second Mrs. Tillotson.
Polly.
Seventy-five Brooke Street.

BY ALBANY DE FONBLANQUE.
Filthy Lucre.

BY R. E. FRANCILLON.
Olympia. | Queen Cophetua.
One by One.

Prefaced by Sir H. BARTLE FRERE.
Pandurang Hari.

BY HAIN FRISWELL.
One of Two.

BY EDWARD GARRETT.
The Capel Girls.

BY CHARLES GIBBON.
Robin Gray.	Queen of the Mea-
For Lack of Gold.	dow.
What will the	In Pastures Green
World Say?	The Flower of the
In Honour Bound.	Forest.
The Dead Heart.	A Heart's Problem
In Love and War.	The Braes of Yar-
For the King.	row.

BY WILLIAM GILBERT.
Dr. Austin's Guests.
The Wizard of the Mountain.
James, Duke.

BY JAMES GREENWOOD.
Dick Temple.

BY ANDREW HALLIDAY.
Every-Day Papers.

BY LADY DUFFUS HARDY.
Paul Wynter's Sacrifice.

BY THOMAS HARDY.
Under the Greenwood Tree.

BY JULIAN HAWTHORNE.
Garth.	Sebastian Strome
Ellice Quentin.	Dust.
Prince Saroni's Wife.	

BY SIR ARTHUR HELPS.
Ivan de Biron.

BY TOM HOOD.
A Golden Heart.

BY MRS. GEORGE HOOPER.
The House of Raby.

BY VICTOR HUGO.
The Hunchback of Notre Dame.

CHEAP POPULAR NOVELS, *continued—*

BY MRS. ALFRED HUNT.

Thornicroft's Model.
The Leaden Casket.
Self-Condemned.

BY JEAN INGELOW.

Fated to be Free.

BY HARRIETT JAY.

The Dark Colleen.
The Queen of Connaught.

BY HENRY KINGSLEY.

Oakshott Castle. | Number Seventeen

BY E. LYNN LINTON.

Patricia Kemball.
The Atonement of Leam Dundas.
The World Well Lost.
Under which Lord?
With a Silken Thread.
The Rebel of the Family.
"My Love!"

BY HENRY W. LUCY.

Gideon Fleyce.

BY JUSTIN McCARTHY, M.P.

Dear Lady Disdain.
The Waterdale Neighbours.
My Enemy's Daughter.
A Fair Saxon.
Linley Rochford.
Miss Misanthrope.
Donna Quixote.
The Comet of a Season.

BY GEORGE MACDONALD.

Paul Faber, Surgeon.
Thomas Wingfold, Curate.

BY MRS. MACDONELL.

Quaker Cousins.

BY KATHARINE S. MACQUOID.

The Evil Eye. | Lost Rose.

BY W. H. MALLOCK.

The New Republic.

BY FLORENCE MARRYAT.

Open! Sesame! | A Little Stepson.
A Harvest of Wild | Fighting the Air.
Oats. | Written in Fire.

BY J. MASTERMAN.

Half-a-dozen Daughters.

BY JEAN MIDDLEMASS.

Touch and Go. | Mr. Dorillion.

CHEAP POPULAR NOVELS, *continued—*

BY D. CHRISTIE MURRAY.

A Life's Atonement.
A Model Father.
Joseph's Coat.
Coals of Fire.
By the Gate of the Sea.

BY MRS. OLIPHANT.

Whiteladies.

BY MRS. ROBERT O'REILLY.

Phœbe's Fortunes.

BY OUIDA.

Held in Bondage.
Strathmore.
Chandos.
Under Two Flags.
Idalia.
Cecil Castlemaine.
Tricotrin.
Puck.
Folle Farine.
A Dog of Flanders.
Pascarel.
TwoLittleWooden Shoes.
Signa.
In a Winter City.
Ariadne.
Friendship.
Moths.
Pipistrello.
A Village Commune.
Bimbi.
In Maremma.

BY MARGARET AGNES PAUL.

Gentle and Simple.

BY JAMES PAYN.

Lost Sir Massingberd.
A Perfect Treasure.
Bentinck's Tutor.
Murphy's Master.
A County Family.
At Her Mercy.
A Woman's Vengeance.
Cecil's Tryst.
Clyffards of Clyffe.
The Family Scapegrace.
Foster Brothers.
Found Dead.
Best of Husbands
Walter's Word.
Halves.
Fallen Fortunes.
What He Cost Her
HumorousStories
Gwendoline's Harvest.
Like Father, Like Son.
A Marine Residence.
Married Beneath Him.
Mirk Abbey.
Not Wooed, but Won.
£200 Reward.
Less Black than We're Painted.
By Proxy.
Under One Roof.
High Spirits.
Carlyon's Year.
A Confidential Agent.
Some Private Views.
From Exile.
A Grape from a Thorn.
For Cash Only.

BY EDGAR A. POE.

The Mystery of Marie Roget.

CHEAP POPULAR NOVELS, *continued—*
BY E. C. PRICE.
Valentina.

BY CHARLES READE.
It is Never Too Late to Mend.
Hard Cash.
Peg Woffington.
Christie Johnstone.
Griffith Gaunt.
Put Yourself in His Place.
The Double Marriage.
Love Me Little, Love Me Long.
Foul Play.
The Cloister and the Hearth.
The Course of True Love.
Autobiography of a Thief.
A Terrible Temptation.
The Wandering Heir.
A Simpleton.
A Woman-Hater.
Readiana.

BY MRS. J. H. RIDDELL.
Her Mother's Darling.
Prince of Wales's Garden Party.

BY F. W. ROBINSON.
Women are Strange.

BY BAYLE ST. JOHN.
A Levantine Family.

BY GEORGE AUGUSTUS SALA.
Gaslight and Daylight.

BY JOHN SAUNDERS.
Bound to the Wheel.
One Against the World.
Guy Waterman.
The Lion in the Path.
Two Dreamers.

BY ARTHUR SKETCHLEY.
A Match in the Dark.

BY T. W. SPEIGHT.
The Mysteries of Heron Dyke.

BY R. A. STERNDALE.
The Afghan Knife.

BY R. LOUIS STEVENSON.
New Arabian Nights.

BY BERTHA THOMAS.
Cressida. | Proud Maisie.
The Violin-Player.

BY W. MOY THOMAS.
A Fight for Life.

CHEAP POPULAR NOVELS, *continued—*
BY WALTER THORNBURY.
Tales for the Marines.

BY T. ADOLPHUS TROLLOPE.
Diamond Cut Diamond.

BY ANTHONY TROLLOPE.
The Way We Live Now.
The American Senator.
Frau Frohmann.
Marion Fay.
Kept in the Dark.

By FRANCES ELEANOR TROLLOPE
Like Ships Upon the Sea.

BY MARK TWAIN.
Tom Sawyer.
An Idle Excursion.
A Pleasure Trip on the Continent of Europe.
A Tramp Abroad.
The Stolen White Elephant.

BY SARAH TYTLER.
What She Came Through.
The Bride's Pass.

BY J. S. WINTER.
Cavalry Life. | Regimental Legends

BY LADY WOOD.
Sabina.

BY EDMUND YATES.
Castaway. | The Forlorn Hope.
Land at Last.

ANONYMOUS.
Paul Ferroll.
Why Paul Ferroll Killed his Wife.

Fcap. 8vo, picture covers, 1s. each.
Jeff Briggs's Love Story. By BRET HARTE.
The Twins of Table Mountain. By BRET HARTE.
Mrs. Gainsborough's Diamonds. By JULIAN HAWTHORNE.
Kathleen Mavourneen. By Author of "That Lass o' Lowrie's."
Lindsay's Luck. By the Author of "That Lass o' Lowrie's."
Pretty Polly Pemberton. By the Author of "That Lass o' Lowrie's."
Trooping with Crows. By Mrs. PIRKIS.
The Professor's Wife. By LEONARD GRAHAM.
A Double Bond. By LINDA VILLARI.
Esther's Glove. By R. E. FRANCILLON.
The Garden that Paid the Rent. By TOM JERROLD.

www.ingramcontent.com/pod-product-compliance
Lightning Source LLC
Chambersburg PA
CBHW021821190326
41518CB00007B/694